AF540477

Animal Byproducts and Wool Science

NIPA® GENX ELECTRONIC RESOURCES & SOLUTIONS P. LTD.
New Delhi-110 034

About the Author

Dr. Bijoy Kumar Sarkar is working in the Department of Livestock Products Technology, College of Veterinary Sciences and Animal Husbandry, R. K. Nagar, West Tripura Tripura -799008 since 2011. He obtained B. V. Sc. & A. H. degree from College of Veterinary Sciences and Animal Husbandry, Selesih, Aizawl, Mizoram, India in 2008. He did his M. V. Sc. in Livestock Products Technology from Division of Livestock Products Technology, Indian Veterinary Research Institute Izatnagar, Bareilly, Uttar Pradesh- 243122 in the year 2010. He cleared ICAR-ASRB NET in the major/broad discipline of Animal Production in 2010 and also cleared the All India Examination, ICAR (SRF) in 2016-17. He achieved Ph.D. in Livestock Products Technology from Department of Livestock Products Technology, College of Veterinary Science, AAU, Khanapara Guwahati-781022 in 2019. Presently he is working as Associate Professor in the Department of Livestock Products Technology, College of Veterinary Sciences and Animal Husbandry, R. K. Nagar, West Tripura, Tripura -799008 since 2024.The author is actively engaged in research, extension and teaching in undergraduate level. So far, he has published more than 25 research paper review and popular articles in reputed national and international journals. He has published two books, numerous book chapters, numerous booklets for farmers and also organized several training programmes for farmers. He received international and national awards for his exceptional devotion to the field of research. He has also served as a mentor to numerous entrepreneurs, helping to create many successful business professionals.

Animal Byproducts and Wool Science

Processing, Utilization, and Sustainable Applications

Bijoy Kumar Sarkar, M.V.Sc. & Ph.D
(Livestock Products Technology)
Associate Professor
Department of Livestock Products Technology
College of Veterinary Sciences and Animal Husbandry
R. K. Nagar, West Tripura-799008 (India)

NIPA® GENX ELECTRONIC RESOURCES & SOLUTIONS P. LTD.
New Delhi-110 034

Dedication

This book is dedicated to my dear daughter, **Sharanya Sarkar** whose sacrifices and understanding have been immeasurable, even when it meant sharing less of my time and affection. Her selflessness encouragement, and constant support have been my greatest source of strength, making this book as much hers as it is mine.

NIPA® GENX ELECTRONIC
RESOURCES & SOLUTIONS P. LTD.

101,103, Vikas Surya Plaza, CU Block
L.S.C. Market, Pitam Pura, New Delhi-110 034
Ph : +91-11-43860225, Mob.: +91 9717133558, 9540816132
E-mail: newindiapublishingagency@gmail.com
Website: www.nipaersources.com

Print ISBN: 978-93-58878-03-5
ebook ISBN: 978-93-58874-42-6

Composed and Designed by NIPA®.

Preface

The growing global emphasis on sustainability has brought renewed focus to innovative approaches in managing animal byproducts and wool. Traditionally seen as waste, materials like hides, bones, fats, and wool are now being recognized for their potential to be repurposed into valuable, eco-friendly products. Wool, in particular, is a renewable resource gaining importance for its biodegradability and versatility in sustainable applications.

"Animal Byproducts and Wool Science: Processing, Utilization, and Sustainable Applications" explores transformative practices and research that drive the future of these industries, offering practical solutions for minimizing environmental impact. Through insights from experts in environmental science, material engineering, and sustainable manufacturing, this book highlights the opportunities and challenges in harnessing the full potential of these resources.

This work aims to inspire innovation and spark global dialogue on the role of sustainability in shaping our production and consumption practices, showcasing how waste can be transformed into valuable resources for a more sustainable future.

I am deeply grateful to my parents, Sri Narayan Chandra Sarkar and Mrs. Arati Rani Sarkar, for their unwavering support in creating this book. Thanks to my sister and brother-in-law for their constant encouragement. I also appreciate my teachers, whose guidance fostered my passion. Special thanks to my students of the College of Veterinary Sciences and Animal husbandry, Tripura, for inspiring me in this journey. I am extremely grateful to my student Bashanti Biswas, for assisting me with the numerous illustrations in this book. Lastly, I am thankful to my friends, family members and colleagues for their support during challenging times, keeping me focused and motivated.

Author

Contents

Section-I: Animal Byproducts

1

Importance of Abattoir Derived Byproducts Processing

India has a substantial livestock population, with animal husbandry acting as a supplementary occupation for about 67% of the rural population. Unlike the commercially oriented farming systems found in Western countries, livestock rearing in India is intricately linked to rural life, where most farmers maintain a small number of animals.

Currently, India's livestock population surpasses 535 million, which includes 192.49 million cattle, 109.85 million buffaloes, 74.26 million sheep, 148.88 million goats, 9.06 million pigs, and over 850 million poultry birds. Each year, approximately 100 million animals are slaughtered in nearly 4,000 slaughterhouses, resulting in a significant amount of by-products or offals. In tropical regions, a considerable portion of these by-products deteriorates before they can be utilized, primarily due to a lack of knowledge, skills, and infrastructure.

Enhancing profitability in the livestock sector could be realized through improved utilization of animal by-products and waste. Many slaughterhouses presently do not have sufficient facilities to collect and preserve these by-products in a timely manner. Additionally, retrieving fallen animal carcasses from remote rural areas often presents challenges. The competitiveness of the meat industry can be bolstered by more effective use of by-products. It is estimated that by-products such as skin, pluck, head, feet, and gut can recover approximately 30% and 35% of the live animal value for goats and sheep, respectively. Thus, it is crucial to add value to significant by-products.

By-products of the meat industry encompass all parts produced in slaughterhouses or butcher shops that are not sold directly as meat. While animals are mainly slaughtered for their meat, a range of additional components, known as by-products or offals, are also obtained. These components may be either edible or inedible. Essentially, "by-products" and "offals" denote all valuable materials derived from a slaughtered animal, excluding the primary meat cuts. For example, from a live cattle weighing 200 kg, approximately 70 kg of meat is obtained, while the remaining 130 kg consists of components such as skin, the digestive system, bones, feet, head, and organs including the liver, heart, kidneys, and spleen. These components are categorized as offals or by-products and are typically removed from the carcass during the processing phase. If these by-products are not utilized effectively, it can lead to diminished returns for producers and increased disposal expenses.

Offals encompass all sections of a slaughtered animal that are not marketed as direct meat cuts. The term by-product is related but encompasses a wider range, including internal organs and other non-meat elements. The effective utilization of animal by-products is crucial not only for financial benefits but also for public health considerations. Industries that convert waste into valuable products should be promoted, as they have the potential to expand by efficiently processing both edible and inedible by-products. Relying exclusively on meat sales is often insufficient for ensuring profitability; thus, the use of by-products serves as a significant source of additional revenue.

When animals are slaughtered, they yield meat, which is a vital protein source, but they also generate by-products that account for more than half of the animal's total weight. To maximize the benefits from this, it is imperative to process and utilize these by-products effectively. The amount of by-products is determined by subtracting the dressing percentage from 100. This dressing percentage, which can range from 45% to 70% depending on the species, indicates the proportion of live weight that is converted into meat.

If by-products are not utilized appropriately, producers may experience a loss of income and incur higher costs for waste disposal. In addition to meat, animals offer several other valuable components, including blood, hides or skins, hooves, horns, head, and contents of the digestive tract such as the stomachs (rumen, reticulum, omasum, abomasum in ruminants), intestines, and various glands. These by-products hold significant value across various industries. For instance, partially digested substances from the stomachs and intestines of ruminants can be processed into fertilizers or animal feed. Pharmaceutical firms also derive significant compounds from animal glands for the production of medications. In India, certain proportions of usable meat and by-products are generally obtained from slaughtered animals, offering valuable resources for various industries.

Table 1: Percentages of saleable meat and by-products from slaughtered animals

Animal	Average yield of saleable meat (%)	Quantity of by-product (%)
Cattle	28-33	67-72
Buffalo	27-30	70-73
Sheep	30-31	69-70
Goat	28-30	70-72
Pig	52-55	45-48
Poultry (Broiler)	55-60	40-45

Classification of Offals

Offals can be categorized according to their function and appearance:

1. By function

a) **Edible offals:** These are parts that are consumed as food, influenced by socioeconomic and cultural factors. For instance, while some individuals may eat the skull, feet, and blood, others may deem these offals unacceptable.

b) **Inedible offals:** These components are either utilized for alternative purposes or discarded, as they are not typically ingested as food.

2. By appearance

a) **Red offals:** These parts are characterized by their reddish hue and include the head, heart, liver, lungs, spleen (commonly referred to as melt), sweetbreads, tail, diaphragm (also known as thick skirt), and tongue.

b) **White offals:** These are lighter in color and comprise fat, the many plies (the third stomach), various sections of the intestines, bladder, tripe (which encompasses the weasand and the first, second, and fourth stomachs, along with the rectum), fore feet, and trimmings.

Handling and Storage of Offals

It is crucial to thoroughly inspect offals for any indications of disease or abnormalities prior to further processing. Items such as blood, liver, trachea, oesophagus, spleen, brain, kidneys, and fat must be collected in a hygienic manner. They should be utilized immediately or stored in stainless steel containers at temperatures ranging from 4°C to 7°C. Blood ought to be processed within 4 to 6 hours, while skin and hides should be transported within 8 hours. The pancreas and endocrine glands should be kept on ice and subsequently frozen. Bones must be cleaned of any remaining flesh, broken, boiled, dried, and then stored at room temperature. Intestines and stomachs should be cleaned in the designated cleaning area of the abattoir.

In comparison to carcasses, edible offals have a higher spoilage rate due to their elevated glycogen levels and reduced fat content. Therefore, they should be separated from the carcass and ideally cooked, utilized, or served on the same day as slaughter. Proper handling is vital to avert contamination. Rapid chilling with carbon dioxide or dry ice is effective in preserving these organs, while freezing halts bacterial proliferation. Vacuum packaging can extend the refrigerated shelf life of organs such as the liver, kidneys, and hearts to approximately 14 days; however, it is advisable to utilize them as soon as feasible. For prolonged storage, organs like liver, heart, oxtail, and giblets should be frozen. When required, these should be defrosted in a refrigerator. Sweetbreads and brains should also be frozen but can be thawed using hot water. Kidney, tongue, and tripe are best consumed fresh or within 24 hours of refrigeration. Tongue can be smoked or pickled, lasting around 3 days when smoked and up to 7 days when pickled. Giblets should be utilized within 12 hours of refrigeration.

a) **Edible offals:** Edible offals, constituting 20-30% of an animal's live weight, necessitate meticulous handling and processing:

1. **Liver:** The liver, the largest organ in the body, is extracted during evisceration. It is crucial to detach the diaphragm and blood vessels without harming the gall bladder to avoid bile contamination, which can impart bitterness. After thorough washing, it should be rapidly chilled, packaged, and dispatched to the market. Although it can be frozen, this process alters its texture, rendering it softer. Liver is utilized in loaves, spreads, and sausages but possesses limited binding capacity due to its high collagen content and pronounced colour. It is abundant in vitamin B_{12}, vitamin A, and iron.

2. **Heart:** The heart, a cone-shaped organ located in the chest cavity, is separated from the lungs, with excess fat removed. It should be opened to inspect for clots and soaked in water for 10-15 minutes prior to chilling. Rich in myoglobin, the heart exhibits a robust colour and is firmer than the liver. It is best suited for moist cooking and is used fresh or frozen in sausage production, offering moderate binding capacity.

3. **Tongue:** Tongue is rich in fat, plays a crucial role in grasping food. It may appear white or feature black spots. The outer layer is eliminated through blanching. It is commonly utilized in luncheon meats and exhibits low to medium binding strength. Tongue can be cured, jellied, or boiled prior to consumption.

4. **Kidney:** These paired organs are sourced from livestock, excluding poultry, as their removal poses challenges and they are generally not

regarded as edible due to the presence of uric acid crystals. Cattle kidneys are characterized by their dark brown colour and lobed structure, while sheep kidneys are shaped like beans. Encased in fat, kidneys are often prepared in various dishes such as pies, stews, and casseroles.

5. **Sweetbreads:** These consist of the tender thymus (located in the neck) and pancreas (found in the gut) glands. They should be washed in cold water and stored on trays in a refrigerated environment, as they require chilling, freezing, or precooking shortly after collection due to their perishable nature. Sweetbreads possess low colour and binding capabilities but are high in collagen, and are frequently prepared by buttering, saucing, frying, or broiling.
6. **Tripe:** Tripe refers to the lining of the stomach, primarily the rumen and reticulum in ruminants, as cleaning the omasum is particularly challenging. It is washed at temperatures between 50-55°C using lime water to eliminate the inner lining, and is often cooked with tomato sauce. Tripe has low binding strength and colour but is rich in collagen.
7. **Brains:** Brains are extracted from the skull and may retain their membrane. If not utilized immediately, they should be frozen or precooked. Delicate and soft in texture, brains are typically prepared by frying or scrambling with eggs. They are rich in collagen but have low binding and colour value. In regions impacted by Bovine Spongiform Encephalopathy (BSE), the consumption of brains is prohibited in both human food and animal feed.
8. **Oxtail:** Once the fat is trimmed, oxtail is employed in soups and stocks, prized for its rich flavour. It can be seasoned to improve the quality of the broth.
9. **Meat extract:** The process of obtaining meat extract involves techniques such as pressing, soaking in cold water, or rapidly boiling in hot water to derive a concentrated juice from the meat. Factors such as the type of meat, cut size, fat content, boiling duration, and treatment methods all significantly affect the yield and quality of the meat extract. It is condensed into a semi-solid form with approximately 20% moisture, comprising various organic and inorganic elements.
10. **Trimmings:** Trimmings denote the tissue mass or structures that are attached to primary organs, such as the delicate layer linked to the diaphragmatic muscle and liver. These trimmings are abundant in fat and contain a moderate level of collagen, which may restrict their

application in ground meat products. They generally exhibit a low to medium binding capacity and share a similar colour profile with the primary meat. Trimmings can also originate from regions such as the oesophageal (weasand) surface, cheeks, tongue, head, and lips, and are frequently utilized in sausage production.

11. **Pig tail:** Similar to oxtail, pig tail is also esteemed for its culinary applications. It is usually brined and cured for a duration of two days or employed to create jelly stock, which is a crucial component in the specialty product known as head cheese.

12. **Pig feet (Trotters):** These are meticulously cleaned using brushes and scrubbers to eliminate any remaining dirt or impurities and are sometimes consumed. The hind feet, which contain less muscle, are less frequently utilized but are processed into products such as tid-bits.

13. **Intestines:** These elongated digestive organs serve both edible and industrial purposes following appropriate cleaning. Intestines are primarily composed of smooth muscle and have various applications once processed. They function as a natural edible casing for sausage preparation. Industrially, they are utilized in the manufacture of catgut, sausage casings, and other products.

14. **Testicles:** These male organs are typically sliced, battered, and deep-fried for consumption.

15. **Pork skins:** Pork skins are widely employed in the production of jellied foods and gelatine due to their high fat content, which imparts strong binding properties. Popped pork rinds, including green belly skins, green back skins, green ham skins, and cured or smoked bacon skins, are commonly utilized. These rinds are often incorporated into emulsion-type meat products such as sausages.

16. **Blood:** Blood serves as a key ingredient in the production of sausages. It is collected in a hygienic manner using a hollow knife, with anticoagulants added to inhibit clotting. Subsequently, the blood undergoes defibrination. Due to its capacity to enhance colour, it is typically incorporated in small quantities, generally ranging from 0.5% to 2%.

17. **Spleen:** The spleen functions as a vital lymphatic organ within the reticulo-endothelial system. It possesses a significant amount of collagen and exhibits strong colour characteristics. The spleen can be fried and utilized in various dishes or incorporated into products such as blood sausage. Nevertheless, the high collagen content may result

in a gritty texture in sausages, potentially limiting its application. Additionally, it is employed in pies or as a flavour enhancer.

18. **Poultry giblets:** Poultry giblets comprise the heart, liver, and gizzard. The heart is trimmed, the gall bladder is extracted from the liver, and the gizzard is opened to remove its contents. Following thorough washing and chilling, the giblets are packaged in a plastic pouch and placed within the cavity of a dressed chicken. Ground giblets are also frequently utilized in soups.

b) **Inedible offals:** The differentiation between edible and inedible offals varies significantly based on cultural practices and economic circumstances. Typically, inedible offals encompass hides and skin, ears, snouts, lips, gall bladders, fetuses, hooves, horns, hair, bristles, etc. In developed nations, inedible offals additionally include items such as blood, feet, heads, lungs, windpipes, spleens, intestines, genitals, trimmings, udders, and all varieties of tripe. However, many of these offals are regarded as edible in less developed countries.

Condemned carcasses, parts of carcasses, and condemned organs are processed alongside other inedible offals. Hides, skins, and bones also represent significant inedible by-products. These inedible materials are transformed into stock feed and fertilizers through the rendering process. The production of stock feed is a crucial method for economically utilizing slaughterhouse by-products. This process entails the transformation of materials such as animals deemed unsuitable for human consumption, deceased animals, bones, discarded hides and skins, tannery remnants, and other waste products into animal feed.

Classification of By-Products

By-products derived from livestock can be categorized based on their use as food, their origin, or their final application.

1. **Based on food use:** Animal by-products are generally divided into two primary categories: edible and inedible. However, this classification is not always clear-cut, as it is influenced by cultural and economic considerations. Factors such as income levels, dietary practices, religious beliefs, and local traditions all play a role in how by-products are categorized. For instance, in affluent, developed nations, items like blood, feet, heads, lungs, windpipes, spleens, intestines, trimmings, udders, foetuses, and tripe are frequently regarded as inedible. Conversely, in numerous developing or less affluent countries, these same by-products are often viewed as edible. In general, animal by-products can be broadly classified into the following categories:

a) **Edible by-products:** These include organs and parts of the animal that are typically consumed as food, although this can vary. Examples encompass kidneys, brain, liver, heart, tongue, oxtail, intestinal tract, gut, and sweetbreads (thymus).

b) **Inedible by-products:** These consist of parts or entire animals that are not suitable for human consumption. This category also includes animals that died prior to slaughter or those that do not pass meat inspection. Examples include ears, lips, snout (the mouth part of a pig), teeth, foetus, gall bladder, trimmings, fleshing, hooves, horns, hair, bristles, hide, and skin.

Certain animal by-products exist in a grey area between edible and inedible, and their classification often hinges on factors such as the animal's health status or contamination during slaughter and processing. Examples in this ambiguous category include the uterus, spleen, testicles, lungs, and blood.

In the United States, the phrase "variety meats" denotes organs that are processed in hygienic conditions. This category encompasses liver, kidneys, heart, oxtail, tripe (stomach), chitterlings (intestines), and fries (testicles).

Numerous by-products, including fat (both consumable and non-consumable), stomachs and intestines, blood, bones, horns, hooves, hair, bristles, hides and skins (for tanning), glands, as well as condemned carcasses or offals, must undergo processing prior to being utilized for their designated purposes.

Table 2: Average breakup of percent yield of different products from food animals

Particulars	**Cattle (% yield)**	**Lamb (% yield)**	**Pig (% yield)**
Carcass and other edible offals	62-64	62-64	75-80
Edible raw fat	3-4	5-6	10-15
Blood	3-4	3.5-4	3-4
Inedible raw material	8-10	6-7	6-8
Stomach and intestinal content	8	5.5	3-4
Hide/skin (pelt and wool in lamb)	7	15	-

Table 3: Percentage yield of meat and different by-products from slaughtered cattle

Name of the products/by-products	**Percentage yield on live weight**
Meat	35
Bones (green)	25
Hide	6.5
Blood	4
Horns and Hooves	0.6
Small intestines	1
Large intestines	1
Stomach	2
Rumen ingesta	11
Liver	1.2
Heart	0.35
Kidneys	0.2
Lungs	1.2
Brain	0.17
Spleen	0.15
Pancreas	0.07
Bile	0.06
Tongue	0.3
Hair, teeth and other	1.5
Cutting losses etc.	8.7
Total	**100**

2. **Based on their origin:** By-products intended for commercial use are generally divided into two principal categories according to their source:
   a) **Primary by-products:** These refer to the initial by-products obtained immediately following slaughter or during the processing phase. They encompass skin, horns, hooves, intestines, feathers, bristles, and feet.
   b) **Secondary by-products:** These are products derived from the primary by-products through additional processing. Examples include leather produced from skin, brushes made from bristles, soup created from feet, neat's foot oil extracted from hooves, artifacts and combs fashioned from horns, and casings derived from intestines.

Table 4: Important animal by-products obtained along with their secondary by-products

Primary by-product	Secondary by-product
Hide/skin	Leather products such as shoes, gloves, belts, bags etc. Collagen sheets, gelatine and glue Stock feed and fertilizer Tallow/ grease
Hide/skin trimmings	Collagen sheets, gelatine and glue Tallow/ grease Stock feed and fertilizer
Bones	Bone meal, gelatine, glue, bone protein, fat, buttons etc. Cutlery handles and bone articles Calcium and vitamin D supplement like ostocalcium tablets
Bristles	Brushes
Wool	Warm garments, lanolin, carpet, fertilizer
Feathers	Feather meal, decorative items, shuttle cock
Blood	Blood meal, albumen, haemoglobin, serum, plasma, fibrin
Intestine	Sausage casings, instrument strings, surgical sutures, tennis racket guts
Horns	Keratin, artifacts
Hooves	Keratin, hoof meal, neat's foot oil
Intestinal contents	Manure
Condemned offals, meat and carcasses	Meat cum bone meal, offal meal, Technical fat or tallow for soap, machine oil, leather dressing, candle etc.
Spleen	Melt
Pancreas	Insulin, pancreatin, glucagon, trypsin
Lung	Heparin, peptone
Liver	Liver extract
Edible offals (lung, liver, spleen etc.)	Pet foods
Stomach	Trips, pepsin, heparin
Calf stomach	Rennin, rennet
Brain	Cerebrosides, cholesterol
Poultry combs	Hyaluronidase
Fat (tallow, lard)	Soap industry, textile industry
Gall bladder	Bile salt, gall stones
Pineal gland	Melatonin hormone
Pituitary gland	GH, LH, FSH, prolactin, oxytocin, vasopressin
Adrenal gland	Adrenalin
Thyroid gland	Thyroxin
Parathyroid	Parathormone

Testes	Testosterone, hyaluronidase
Ruminal and intestinal ingesta	Stock feed, compost manure, Production of methane for light, heat and power

3. **According to ultimate use:** By-products can be categorized based on their final application into three primary types:

   a) **Agricultural by-products:** These are mainly utilized in farming and agriculture. Examples include:

   - **Meat meal**: Employed as animal feed.
   - **Bone meal:** Utilized as a fertilizer.
   - **Fertilizer:** Produced from various animal by-products to improve soil quality.

   b) **Industrial by-products:** These are applied in a range of industrial uses. Examples include:

   - **Gelatine:** Incorporated in food products, pharmaceuticals, and cosmetics.
   - **Glue:** Created from collagen, used as adhesives.
   - **Casings:** Employed in sausage production and other food processing.

   c) **Pharmaceutical by-products:** These are utilized in the manufacturing of medications and health-related products. Examples include:

   - **Insulin:** Used in the treatment of diabetes.
   - **Pepsin:** Utilized in digestive aids and enzyme therapies.
   - **Bio-chemicals:** Various chemicals employed in medical research and treatments.
   - **Hormones:** Used in hormone replacement therapies and other medical applications.

Benefits From Utilization of Animal By-Products

By-products constitute approximately 50% of an animal's live weight, thus their efficient use can greatly improve returns for producers. Effective management and utilization of by-products also contribute to:

1. **Improved environmental sanitation:** By-products such as blood, meat trimmings, fleshing, condemned organs, and unused offals can attract pests like flies, dogs, vultures, and jackals, leading to public nuisance and heightened disease risk. These materials decompose rapidly, generating unpleasant odours and creating an environment conducive to harmful microbial growth. Handling meat under these unsanitary conditions diminishes its quality and poses health risks. Timely and

proper processing of by-products, along with the establishment of dedicated by-product facilities, enhances slaughterhouse hygiene and aids in reducing environmental pollution.

2. **Healthier livestock and improved soil fertility:** The strategic utilization of animal by-products plays a significant role in enhancing both livestock health and soil fertility.
   a) **Bone meal:** When mixed with salt and trace minerals such as copper (Cu), iodine (I), cobalt (Co), and iron (Fe), bone meal acts as a comprehensive mineral supplement for livestock.
   b) **Blood meal:** This by-product is abundant in protein, containing over 80% protein content, thus serving as a valuable addition to carbohydrate-rich feeds.
   c) **Meat and bone meal:** This by-product not only provides high-quality protein but also essential nutrients like phosphorus (P), calcium (Ca), and vitamin B_{12}, which are vital for animal growth.
   d) **Bone meal for soil fertility:** The incorporation of bone meal into soil significantly enhances its fertility, resulting in improved crop yields and overall soil health.
3. **Promotion of secondary rural industries:** The processing of secondary by-products from animals stimulates the development of rural industries:
   a) **Tanning industry:** High-quality hides and skins are the foundation of a prosperous tanning industry.
   b) **Soap production:** Tallow and other animal fats are utilized locally in the manufacturing of soap.
   c) **Brush manufacturing:** Hair and bristles are employed in the creation of various types of brushes.
   d) **Animal feed:** Inedible fat can be integrated into high-calorie diets for broilers, thereby supporting poultry nutrition.
   e) **Casing production:** The utilization of animal casings contributes to the growth of secondary rural industries, particularly in the production of sausages and similar products.
4. **Impact on meat price structure:** The appropriate use of by-products can lead to a reduction in meat production costs and influence meat prices, ultimately resulting in enhanced profits for livestock farmers. By-products such as blood, bones, fat, hides, skins, intestines, and stomach contents can generate substantial income for processors, thereby increasing overall profitability and potentially decreasing meat production expenses.

5. **Generation of employment opportunities:** The effective transformation of inedible offals into valuable by-products generates job opportunities and fosters skill development. In contrast to basic waste disposal, which requires minimal labour, the conversion of these by-products into useful products creates employment and bolsters industries reliant on these materials. This not only benefits the primary production sectors but also stimulates economic growth in associated industries.

Status of By-Product Industry

1. **Developed countries:** In developed nations, animal slaughter is centralized and conducted in well-regulated abattoirs. This systematic approach yields a substantial and consistent supply of both edible and inedible offals. Such a framework renders the processing of by-products economically viable and efficient, guaranteeing their appropriate utilization and processing.
2. **Developing countries:** In developing nations, where approximately 80% of the population resides in rural areas, meat is typically sourced from freshly slaughtered animals. Roughly 80-85% of slaughtering occurs in these rural locales, often through informal or unregulated practices. Due to lower purchasing power and a reduced number of animals being slaughtered, the availability of edible and inedible offals is limited, resulting in less profitable processing opportunities. India is experiencing growth in modern abattoirs and meat processing facilities that incorporate by-product processing capabilities. In rural areas, individuals engaged in by-product processing generally gather skins and intestines from various sources and process them manually. Conversely, urban processors obtain raw by-products from slaughterhouses and manage them based on their available resources and facilities.

2

Plan and Layout of Byproduct Utilization Plant

Given our substantial livestock population, it is essential to efficiently process animal by-products to enhance economic gains and generate employment opportunities for individuals with limited educational backgrounds. To effectively manage the high density of livestock, it is advisable to establish animal by-product processing facilities at intervals of every 50 kilometers. Additionally, situating these facilities in proximity to abattoirs can contribute to the reduction of pollution. Furthermore, the waste generated from both abattoirs and processing plants can be utilized to produce biogas, offering mutual benefits.

Incorporating by-product utilization facilities is a critical component of modern abattoirs. This approach not only boosts profits for producers but also mitigates environmental pollution resulting from improper waste management. In nations such as India, where there is a scarcity of modern abattoirs and many operate using outdated practices, it becomes particularly vital to establish facilities dedicated to the management of deceased animals and slaughterhouse waste. Some developing countries have even resorted to employing mobile units to effectively manage dead and fallen animals.

While it is frequently asserted that every part of an animal, with the exception of its last cry, holds value when utilized appropriately, the actual utility of by-products is contingent upon various factors, including the size of the abattoir (the number of animals processed daily), the availability of raw materials, the current applications of by-products, and their significance.

A standard by-product utilization facility encompasses various components such as rendering plants, hide/skin storage, and casing processing units. These facilities gather raw by-products and transport them to different processing centers located away from the slaughterhouse. For example, glands may be directed to the pharmaceutical sector, skins to tanning operations, fat to soap production, and bones to gelatine extraction. When designing such a facility, it is crucial to consider the availability and quantity of raw materials to ensure efficient functioning and comprehensive utilization of by-products.

General Considerations

When setting up a byproducts utilization facility, several critical factors must be taken into account to guarantee effective and efficient functioning.

1. **Integration with slaughterhouse:** A byproducts facility can be incorporated into the slaughterhouse structure, provided that it is distinctly separated from the slaughter activities to prevent contamination.
2. **Legislative restrictions:** Certain regions mandate that the facility be distinct from the abattoir. Compliance with local laws and regulations is crucial.
3. **Proximity and access:** Ideally, the byproducts facility should be situated close to the slaughterhouse but not within the same building. It should feature a separate entrance for the delivery of raw materials.
4. **Location considerations:** It is advisable to establish the combined slaughterhouse and byproducts facility in less densely populated regions to lessen the impact on local communities.
5. **Concrete passageway:** If the slaughterhouse and byproducts facility are adjacent, the passageway between them should be paved with concrete, stones, or bricks, and sloped towards the byproducts facility to avert pollution and drainage complications.
6. **Overhead rail system:** An overhead rail system for transporting condemned carcasses and offals in leak-proof containers directly from the abattoir to the byproducts facility can enhance efficiency.
7. **Transportation vehicles:** Trucks utilized for transporting carcasses or offals should be constructed of metal to uphold hygiene and equipped with rubber wheels to minimize damage to floors and pavements.
8. **Drains and grease traps:** The byproducts facility should possess its own screened drains leading to a gutter with a separate grease trap.
9. **Drainage system:** All drains should be directed away from the slaughter hall to prevent contamination.
10. **Plant layout:** The facility must be organized into distinct clean and unclean areas. The unclean section is designated for receiving raw materials, while the clean section is allocated for processing and sterilizing by-products. Each section should have its own entrance and exit.
11. **Section isolation:** It is imperative that there is no interaction between the clean and unclean sections of the facility; each section must possess its own access points.

12. **Construction materials:** The floors, walls, and ceilings should be constructed from smooth, concrete materials to enable easy and regular cleaning. A recommended floor slope is half an inch per foot.
13. **Ventilation:** The structure should be adequately ventilated to mitigate humidity, which can lead to corrosion and mould proliferation. Corrugated asbestos sheeting is suggested as a roofing material, and the installation of exhaust fans is advised.
14. **Essential facilities:** The facility should encompass a hide/skin salting room, renderers, tripery, boiler, laboratories, manure bunker, and storage areas, in addition to standard amenities.
15. **Equipment:** The equipment must be designed to manage the anticipated daily volume of byproducts or offal. Key equipment includes renderers, fat settling tanks, fat expellers, blood presses, and grind mills. A boiler with a steam pressure capacity exceeding 80 PSI is essential for processing and cleaning.
16. **Phased implementation:** Initiate the process by handling only one or two types of by-products, and progressively expand operations in accordance with demand and capacity. Addressing these factors will facilitate the establishment of a byproducts utilization plant that functions efficiently, adheres to regulations, and satisfies hygiene and environmental standards.

Plant Layout

From a hygiene standpoint, akin to the slaughterhouse, a by-products plant must ensure a clear separation between areas that manage raw materials and those that handle processed products. Given that one of the end products is meat and bone meal for livestock feed, it is crucial to design and operate the plant in a manner that prevents any contact between the clean and unclean zones.

1. **Unclean sections:** The unclean area, where infectious materials are introduced, comprises the following components:
   a) Unloading platform.
   b) Skinning and cutting room designated for deceased animals.
   c) Room for salting and storing hides and skins.
   d) Gut processing area equipped with cleaning tables, water taps, etc.
   e) Carcass utilization facility featuring a renderer, bone digester, fat settling tank, fat expeller, along with an adjacent milling room and storage area.

f) Cloakrooms, washrooms, and a dining area for the staff involved, as well as for the truck drivers collecting offals and deceased animals.

The unclean section of the by-products facility should be fitted with smooth concrete or tiled floors and walls to facilitate efficient cleaning and disinfection. For hygiene purposes, workers must have access to facilities for washing and changing their clothing post-shift. This includes a cloakroom with distinct sections for work attire and personal clothing, as well as a washroom equipped with both hot and cold showers.

The hide salting and storage area must also be designed for easy cleaning, featuring tiled surfaces and adequate drainage. Hides should be salted and stored for a minimum of 14 days prior to being sent to the tannery to mitigate the risk of infection. To prevent contamination, animals and birds must be excluded from this area.

Condemned meat, offals, and deceased animals should be transported in specially designed vehicles that are easy to clean and disinfect. These trucks must have waterproof flooring to avert leakage of infectious materials during transit.

2. **Clean sections:** The clean area, where sterilized materials are processed and stored, includes the following:
   a) Room for sterilization equipment.
   b) Room for hydraulic press and/or centrifuge.
   c) Room for a milling unit.
   d) Room for fat refining.
   e) Rooms designated for the storage of meat meal, meat and bone meal, and technical fat.
   f) Washroom, cloakroom, and dining area for the personnel.

The floors and walls of these rooms should be constructed from materials that facilitate easy cleaning and disinfection. An autoclave designed for the sterilization of manure and wastewater must be linked to the clean section of the facility to guarantee effective treatment.

To uphold a strict division between clean and unclean zones, the arrangement of equipment should ensure that the renderer's charging hopper is situated in the unclean area, while sterilized products are discharged through a door into the clean section, where milling, packing, and dispatch occur. This separation is reinforced by a partition wall constructed across the melter, which prevents any direct interaction between the two areas.

The layout of the plant should promote a continuous and seamless flow of operations among the equipment, thereby minimizing the potential for airborne contamination. Whenever possible, covered screw conveyors ought to be employed to transfer materials between processing locations to uphold hygiene standards.

Floors must be inclined at a minimum of half an inch per foot to guarantee adequate drainage. Standard provisions for hot water and steam cleaning, including hosepipes, are crucial for effective sanitation. The boiler should be appropriately sized based on the volume of raw materials and renderer usage, with an estimated steam requirement of around 1.25 pounds for each pound of raw material processed.

This design and operational configuration significantly mitigates the risk of contamination post-processing, ensuring that the final product is both safe and of superior quality. The various sections within the by-products plant should be organized as follows:

a) Rendering unit
b) Gut processing unit
c) Hide/skin store
d) Horn/hoof processing unit
e) Gland collection and primary collection unit
f) Store
g) Effluent treatment plant
h) Electrical room
i) Boiler unit
j) Mechanical section
k) Refrigeration unit
l) Office
m) Toilet, bath and changing rooms
n) Despatch section

3. **Ventilation:** Rendering operations generate a significant amount of steam, which, if not adequately controlled, may result in corrosion of both the building and its equipment. Excessive moisture also promotes the growth of mould and bacteria, while insufficient ventilation leads to uncomfortable working environments. To avert these issues, it is essential for the building to have proper ventilation. A simple yet effective approach is to implement a prefabricated steel structure that features a ventilating ridge at the apex of the roof.

4. **Deodorisation:** Rendering plants frequently emit unpleasant gases as a result of processing deceased and decomposed animals. To effectively manage and mitigate these odours, various strategies can be utilized:

a) **Burning fumes:** Channel the fumes from the rendering vessel into the boiler stack or combustion chamber, where they can be incinerated and dispersed.

b) **Condensation:** Introduce hot vapours into cooling water, allowing them to be absorbed and subsequently released into the effluent disposal system. This process can be facilitated by a condenser, some of which utilize a vacuum pump.

c) **Chemical treatment:** Employ chemical techniques such as chlorination or absorption with activated carbon to treat and neutralize the odours. Moreover, dry rendering equipment generally emits less offensive odour in comparison to wet rendering.

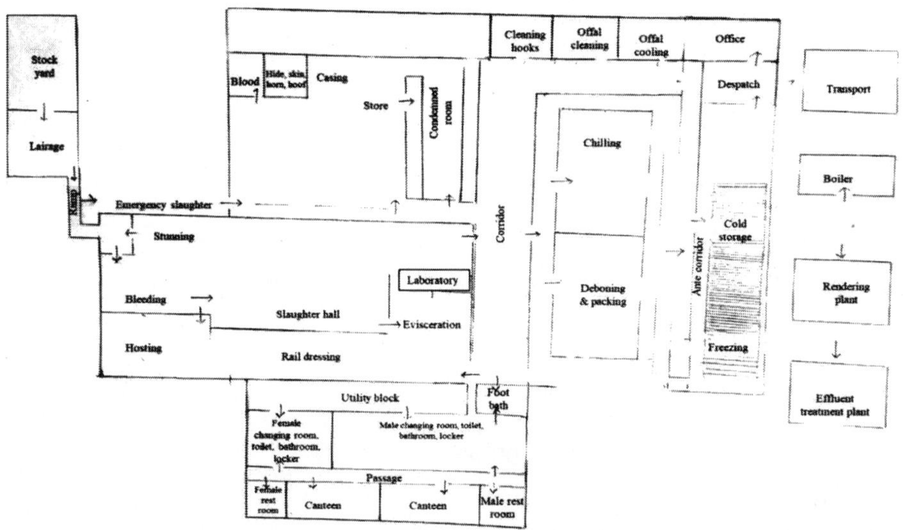

Figure 1: General layout of an abattoir

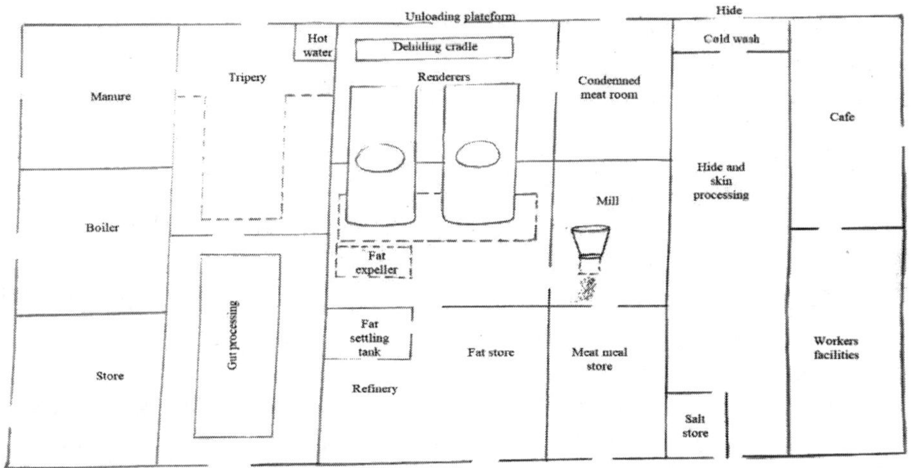

Figure 2: Layout of byproducts processing plant

Equipments

In addition to a boiler, a by-products plant necessitates several essential pieces of equipment, which include a rendering melter or cooker, fat extractors or expellers (such as hydraulic presses), a bone digestor, and a milling unit. These machines are available in various sizes, typically with capacities ranging from half a ton to two tons. Processing offals in batches smaller than one-fourth ton using standard equipment is generally not cost-effective. The equipment must be appropriately sized to accommodate the daily volume of raw materials. Inadequate capacity can result in offal decomposition, leading to unpleasant odours, significant bacterial contamination of processed products, diminished protein yield, and operational challenges.

1. **Boiler:** An adequately sized boiler is crucial for efficient rendering operations, as the average steam consumption is approximately 1.25 pounds of steam per pound of raw material processed. To improve efficiency and decrease rendering time, raw materials should be chopped or minced prior to processing. To reduce heat loss, steam pipelines and the boiler should be insulated with fiberglass and covered with an aluminum sheet. Furthermore, it is vital to utilize a boiler capable of sustaining a working steam pressure of at least 80 pounds per square inch (PSI), especially when producing by-products such as meat meal and bone meal.
2. **Wet renderer:** Wet rendering entails processing raw materials with the addition of water or steam condensate. Steam comes into direct contact with the material inside a vertical cylindrical cooker, which typically

features a cone-shaped bottom and an 8 to 12-inch diameter valve outlet. The top of the cooker is equipped with a manhole for loading and a valve to release gases without disrupting the internal pressure. Fat and water are extracted through side draw-off cocks. Operating at approximately 40 pounds per square inch (PSI), the process usually lasts between 4 to 8 hours. Following cooking, the mixture, referred to as tankage or slush, is allowed to settle for 2 hours. Initially, the floating fat or grease is removed, followed by the extraction of water, after which the remaining tankage, which consists of digested meat and bones, is gathered. This mixture retains approximately 55% moisture and necessitates additional drying to lower the moisture content to below 8%. The bones that are separated during this procedure are directed to a bone digester. Wet rendering typically results in a higher fat recovery compared to alternative rendering techniques.

3. **Dry renderer:** Dry rendering eliminates moisture from raw materials while maintaining essential nutrients. It employs a horizontal steam-jacketed cooker equipped with central agitators that continuously mix the material, ensuring uniform heating and preventing burning. In contrast to wet rendering, steam only heats the jacket of the cooker and does not come into direct contact with the material. The process functions at 60-80 PSI and usually lasts between 3 to 4 hours, cooking the material in its own moisture. Dry rendering is more effective than wet rendering, producing up to 20% more product because water-soluble proteins and extracts are preserved rather than discarded. This technique effectively cooks and fries meat and bones in their natural fluids and fat, leading to savings in steam and labour. The end product, referred to as cracklings, is gritty, fibrous, and non-slippery, with residual fat that should be reduced to approximately 5% to decrease labour costs and improve storage quality. Long bones are separated from the cracklings and sent to a bone digester. The removal of fat from cracklings can be accomplished through:

a) **Mechanical methods**

- **Hydraulic press:** Applies a pressure of 4000 pounds per square inch.
- **Centrifugal turbine fat extractor:** Utilizes centrifugal force to separate fat.
- **Fat expeller:** Fat expeller extracts the fat. Even with these techniques, around 10% fat remains in the cracklings.

b) **Chemical methods**

- **Solvent extraction:** Employs solvents such as petrol, ether, heptane, and perchloroethylene to reduce fat content to as low as 0.5%.

4. **Bone digester:** Long bones undergo crushing and steam processing at 60 PSI for approximately 2 hours. This procedure effectively separates the majority of the protein and fat from the bones. Following the extraction of fat and gelatin, the remaining digested bones take on a chalky and soft texture, facilitating their grinding into bone meal. The resultant bone meal generally comprises about 32.5% calcium and 15% phosphorus.
5. **Milling unit:** Once the fat is extracted from the cracklings, they are processed into a fine powder utilizing machinery such as grinders, disintegrators, or swing beaters. The resulting powdered product is subsequently collected via a cyclone and sacking unit, enabling it to be packaged directly from the milling operation.

Table 5: Technical details of equipment required for carcass utilization plants

Sr. No.	Item description	Use
1.	De-hiding cradles	For de-hiding and removal of skin.
2.	Containers for offal	For carrying offals.
3.	Pre-breaker	For breaking large animal carcasses (up to 500 kg) into smaller particles.
4.	Dry melter (dry rendering cooker)	For cooking the material at high temperature and pressure.
5.	Charging chute	For charging the dry melter with raw material.
6.	Percolating tank	For collecting digested material and for drainage of free fat from the greaves.
7.	Fat transfer tank unit	For fat transport to settling tank.
8.	Fat settling tank (1000 litters)	For collecting hot fat from the percolating tank and from the centrifugal fat extractor.
9.	Centrifugal fat extractor	For centrifugation of fat and for removal of clean fat. The fat content of the greases should be brought down to a minimum of 8-12 percent after centrifugation.
10.	Cooling platform	For cooling defatted greaves.
11.	Hammer mill	This is to reduce particle size. It should have a bagging spout and bag holder. Provision of stone trap in the unit. Automatic bagging unit should include a balance hanger for setting specified load.
12.	Inspection/operation platform	Fabricated from mild steel, inclusive of handrail and staircase. Placed beside the cooker and pre-breaker. Size: 2 meter x 2 meter, 1.5 meter height.

13.	Steam boiler complete with water softener unit and PWPS (Plant Water Purification System)	For supply of steam for cooking. Supplied as a complete unit with capacity 600 kg/hr for the mentioned plant. The boiler will automatically cut off at high temperature, low water level and high pressure.
14.	Shell and tube condenser	To minimize odour.
15.	Non-condensable gas tank	For removal of gases to chimney.
16.	Cyclone	For separation of particles from the vapour.
17.	Cooling tower	For cooling.
18.	Water circulating pump	For circulating the water through condenser and cooling tower.
19.	Chain hoist, capacity 500 kg and motorized trolley	For lifting carcass.
20.	Electric control panel and switch panel	For controlling various operations.
21.	Automatic moisture control	Control unit with instrument and stabilized DC supply unit for the measuring circuits. The module accommodates two data display for count adjusts and maximum count set.
22.	Set of knives with wooden handle Comprising of: 10 nos. De-hiding knives 10 nos. Choppers 10 nos. Butchers knives	For cutting carcass.
23.	Set of spare and wear parts	Recommended spare and wear parts for 3 years operation.
24.	Set of tools	For operation or maintenance of all the equipment and machinery of the plant. The complete list of tools should be provided item wise.
25.	Motorized multipurpose saw	For cutting carcasses.
26.	Effluent treatment plant	For treating the effluents arising from the carcass in the plant. The BOD (Biological Oxygen Demand) reduction should be to level of 250-300 mg per litre.

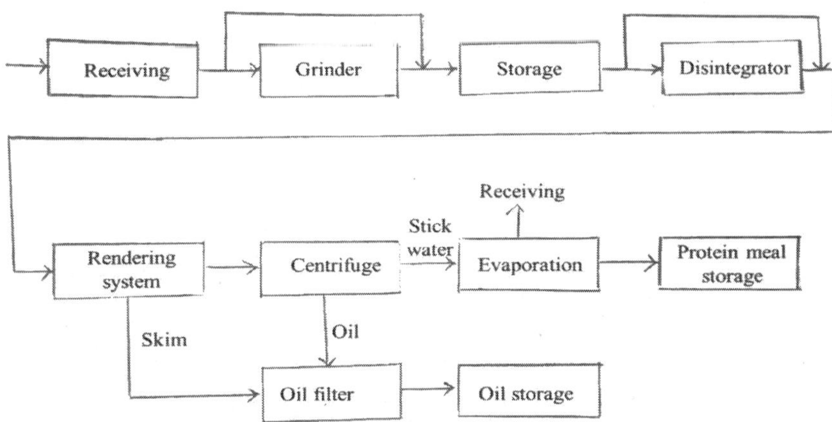

Figure 3: Basic operations of byproducts processing plant

Handling of Dead or Fallen Animals

1. **Transportation of dead or fallen animals:** In both developing and underdeveloped nations, numerous animals succumb to natural causes, including diseases. In India, the prohibition of cow slaughter for consumption further exacerbates the number of deceased animals. Moreover, condemned materials originating from slaughterhouses add to the quantity of available byproducts. It is crucial to ensure the hygienic disposal of these fallen animals to avert decomposition and the emission of harmful gases. The processing of these byproducts not only alleviates environmental risks but also yields economic advantages.

   Transportation of deceased animals is generally performed through terrestrial transport means, including bullock carts or tractor trolleys. Bullock carts are generally employed for shorter distances due to their adaptability and convenience, whereas tractor trolleys are favored for longer journeys owing to their enhanced capacity and speed. To maintain the integrity of hides and skins, the floors of vehicles should be lined with straw and devoid of any sharp edges that could inflict damage. When moving condemned materials from slaughterhouses to byproduct facilities, it is imperative to utilize closed vehicles that are distinctly marked as transporting inedible animal byproducts, thereby ensuring hygiene and safety throughout the operation.

2. **The utilization of deceased or fallen animals:** The hides and skins of deceased animals represent valuable raw materials utilized in the manufacture of leather and various leather products. The carcasses of deceased or fallen animals are processed at a carcass utilization facility,

which manages various components such as carcass pieces, bones, heads, feet, lungs, trachea, spleen, intestines (cleaned and devoid of contents), trimmings, genitals, udders, fetuses, and occasionally tripe. However, certain materials such as hooves, horns, hair, blood, and stomach or intestinal contents are excluded from this procedure. Instead, these excluded components are typically repurposed in the production of fertilizers and other byproducts.

3. **Preparation of meat meal from deceased or fallen animals:** Upon the receipt of fallen carcasses or condemned meat, the hides, horns, hooves, viscera, and gut contents are initially removed. The carcass is subsequently manually disassembled into smaller fragments and fed into a pre-breaker, which reduces the size to approximately 60 mm. These fragments are then placed into a dry renderer, where they undergo cooking and sterilization at 75 PSI for a duration of 3 to 4 hours. Following the cooking process, the resultant material, referred to as grieves, is transferred to a percolating tank to facilitate the separation of fat from the cracklings.

   The cracklings are subjected to further processing utilizing a centrifugal fat extractor to eliminate excess fat and are pressed to extract additional fat. The defatted material is then ground in a milling unit to create meat meal, which is ultimately packaged and labeled for distribution.

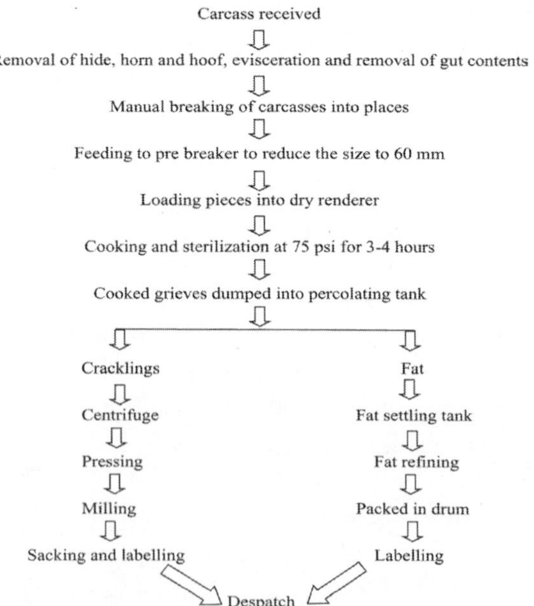

Figure 4: Flow diagram for utilization of dead or fallen animals

3

Treatment of Condemned Meat and Carcasses

Rendering is the process of transforming deceased animals or condemned meat into valuable products, such as carcass meal. Traditionally, rendering specifically referred to the extraction of fat from animal materials through heating. However, the term has evolved to encompass the production of carcass meal, meat meal, and technical fat. While the traditional method of rendering involved boiling animal remnants to extract fat, contemporary rendering includes the heating or steaming of carcasses to produce a nearly sterilized material that retains nutrients. The animal by-products utilized in rendering consist of excess fat, hooves, and offal. In the past, tallow was employed in the manufacture of soap and candles, while other by-products like heads and hooves were repurposed for fertilizers. Meat packers quickly recognized the potential for profit in processing these by-products. Today, the rendering industry generates a variety of valuable products, including both edible and inedible oils, chemicals, meat meals, and bone meals, derived from materials such as viscera, bones, trimmings, dead stock, and feathers. A substantial portion of these by-products is recycled into useful products. In developed nations, as well as in India, integrated abattoirs and meat processing facilities manage such waste utilizing advanced equipment and sophisticated processes.

Pre-rendering operations: It is essential to process raw materials from the slaughter floor swiftly and comprehensively to guarantee the highest quality of the final product. Prior to placing these materials into the cooker, they require adequate preparation. Heads, feet, condemned carcasses, and bones from the boning department are initially reduced or crushed into smaller fragments using shredders. Additionally, bones can be further reduced in size with machines referred to as "pre-breakers." Intestines and other soft tissues, which may be contaminated with feed or manure, must be opened, cleaned, and reduced in size before being incorporated into the cookers during the inedible rendering process.

Preparation of carcass meal: The preparation of meat and bone meal, or carcass meal, involves thermal treatment processes such as rendering or sterilization. There are two primary technologies for processing carcasses: dry rendering and wet rendering. Continuous low-temperature dry rendering is particularly effective in yielding higher-quality fats. For large animals such as cattle and buffaloes, the value of their by-products encompasses not only their hides, hooves, and horns but also products like meat meal, bone meal, and tallow. Once the hide, horns, and hooves are removed, the remaining portions of the carcass may either be discarded or subjected to further processing. In a carcass rendering facility, the feedstock generally excludes hides, horns, hooves, and rumen contents. This material, which contains varying proportions of protein, fat, and water, is rendered to extract proteins and fats while eliminating water. In the absence of rendering, the material would undergo decomposition. The yield of final products such as meat meal, bone meal, and tallow are significantly influenced by the weight and nutritional quality of the animal.

Rendering Systems

In general, there are following four categories of rendering system.

a) **Batch wet rendering or Autoclave rendering:** In a fundamental autoclave or batch wet rendering system, the procedure initiates with the autoclave or cooker being filled with pre-ground raw material. Subsequently, the autoclave is sealed, and steam is introduced at a temperature of approximately 110°C and a pressure of 40 PSI. This treatment endures for a duration of 4 to 8 hours. Upon completion of the treatment, the pressure is decreased to atmospheric levels. Thereafter, the material is extracted from the cooker. The free-floating fat is drained, and the residual moist crackling or solid material is pressed to eliminate further liquid before undergoing the drying process. The liquid obtained from this procedure comprises fat, water, and fine particles. The fat is then isolated from this mixture through either gravity separation or centrifugation. The production of bone meal predominantly depends on this wet rendering technique.

b) **Batch dry rendering:** The dry rendering technique utilizes heat in the form of steam and water, albeit indirectly. This method requires maintaining a pressure of 60-80 PSI (with an average of 75 PSI) for a duration of 3-4 hours to facilitate the evaporation of water from the fat within the cooker, thereby preserving essential nutrients. In contrast to wet rendering, steam and hot water do not directly enter the cooking vessel; rather, steam is introduced into the outer jacket of the cooker. The raw material, which is ground to a size of less than 2.5 cm, is batch-fed into the sealed cooker. Typically, dry rendering produces approximately 20% more product than wet rendering.

c) **Continuous dry rendering system:** This approach is akin to the batch dry rendering system, wherein moisture is eliminated through the application of dry heat. The primary distinction is found in the manner in which the material flows into and out of the cooker. In this system, the process encompasses:

- **Material flow into the cooker:** The raw material is continuously batch-fed into the cooker.
- **Moisture removal:** Moisture is expelled using dry heat, which entails heating the material without the direct introduction of steam or water into the cooking vessel.
- **Material flow out of the cooker:** Following the rendering process, the material is extracted from the cooker. The application of dry heat guarantees effective moisture removal and can result in higher yields when compared to alternative rendering methods.

d) **Continuous low temperature system:** This method is referred to as the mechanical dewatering system and bears resemblance to the conventional batch wet (autoclave) rendering system.

- **Mechanical dewatering:** Its main objective is to eliminate moisture from the material through mechanical means instead of direct contact with steam or water.
- **Process similarity:** Similar to the traditional batch wet rendering system, the mechanical dewatering system incorporates heating; however, it utilizes dry heat and mechanical pressure to expel moisture, thereby maintaining a comparable overall processing approach.
- **Material handling:** The raw material is processed in batches, and the system guarantees efficient moisture extraction while safeguarding the nutrients. The mechanical dewatering system is engineered to enhance the removal of water from fats and other constituents, potentially resulting in increased yields and greater efficiency.

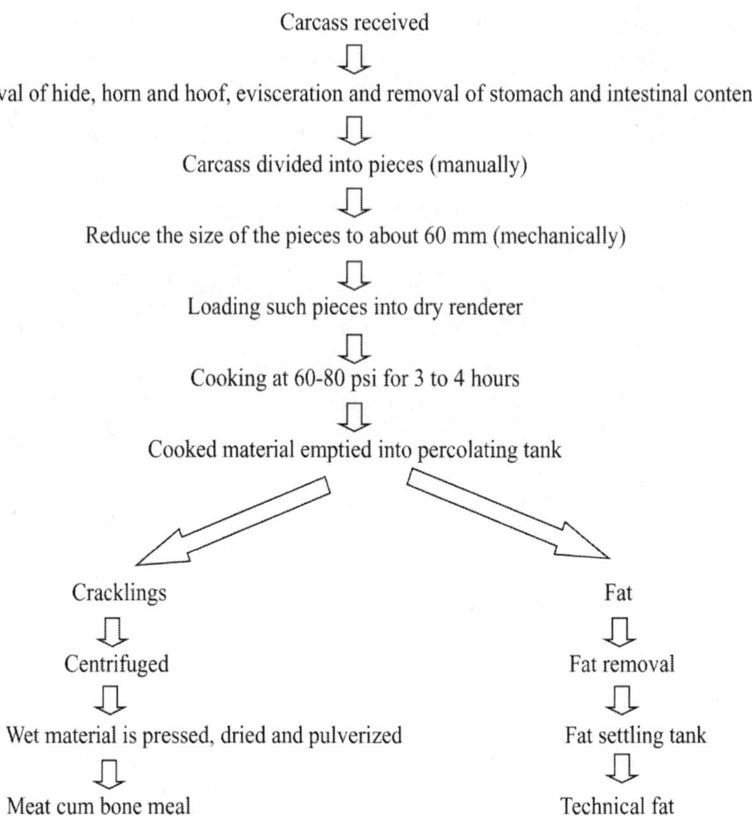

Figure 5: Flow diagram for dry rendering

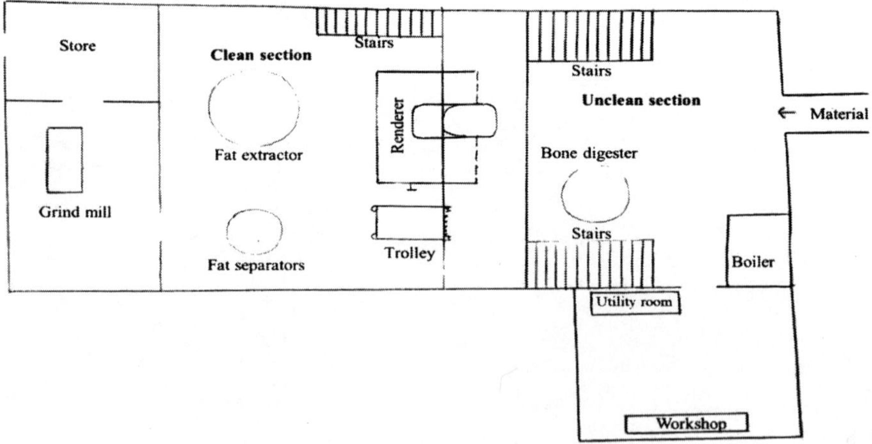

Figure 6: Layout of carcass utilization plant

Table 6: Comparisons between Dry Rendering and Wet Rendering

Wet Rendering	**Dry Rendering**
In wet rendering the material is cooked in added water.	In dry rendering the material is cooked in its own water.
Temperature applied is about 110°C and pressure of 40 psi, for 4-8 hours followed by drying for about 3-4 hours.	Temperature applied is about 130°C and pressure of 60-80 psi, for 3 to 4 hours. Better sterilization of material at higher Temperature.
The cooker is of vertical type.	The cooker is of horizontal type.
The recovery of fat is better than dry rendering. To produce good grade tallow, viscera must be cut and washed.	The recovery of fat is not as effective with wet rendering. Tallow is dark coloured. More tallow is lost in dry rendered meat.
Yield of carcass meal is less than dry rendering as 20-25% of meat is lost in the gravy.	Yield of carcass meal is more (20-25%) than wet rendering.
The resultant material is called slush or tankage.	The resultant material is called cracklings.

Figure 7: Wet renderer

Figure 8: Dry renderer

Main By-Products Obtained By Rendering

Here are the main by-products obtained from rendering:

1. **Tallow/lard:** Tallow is the fat that has been rendered from cattle and sheep, whereas lard pertains to the rendered fat derived from pigs. In comparison to tallow, lard or grease typically exhibits a lower titre value. The extent of fat spoilage is determined by the concentration of

free fatty acids, which should remain below 2%. The hue of fat can differ, ranging from white to yellow, greenish-brown, or red, influenced by factors such as the breed, diet, age, and overall condition of the livestock. The green tint in tallow is often a consequence of contact with intestinal contents. During dry rendering, excessive heat can cause tallow to take on a reddish appearance, while the presence of blood may result in a brownish hue. To preserve the desired colour and quality, it is crucial to utilize fresh, uncontaminated materials. This includes measures such as preventing blood and gut contents from entering cookers and maintaining regulated temperatures and pressures. Pure fat should be almost devoid of moisture, as water in tallow can facilitate bacterial proliferation. Soap manufacturers often bleach tallow to rectify colour discrepancies prior to further processing.

2. **Meat and bone meal:** Meat and bone meals serve as abundant sources of protein, energy, B-complex vitamins, and minerals. They generally contain approximately 50% crude protein and 8 to 12% phosphorus. Adding around 10% meat meal to animal feed fulfills the nutritional needs for essential amino acids, including lysine, methionine, threonine, and tryptophan. The bone content in meat and bone meal provides calcium and phosphorus, with the phosphorus from bones being more accessible to animals compared to some alternative phosphorus sources. Furthermore, meat and bone meal is a significant source of vital B-complex vitamins, such as thiamine. Rendered animal protein, exemplified by meat and bone meal, is frequently utilized in pet food.
3. **Bone meal:** A ground product created by cooking dried bones with steam under pressure to eliminate grease and meat fibers, frequently utilized in the manufacturing of gelatine or glue. It is appropriate for animal feed and comprises up to 15% protein, along with a significant mineral content. Bone meal serves as an essential supplement in animal feed due to its elevated calcium levels, which support bone development in poultry and provide the necessary calcium for egg shells. Additionally, it contributes phosphorus and calcium to livestock and pet foods. Animal glue is derived from collagen extracted from bones by heating them in water. Following the extraction process, the water is evaporated to yield the glue. To produce bone meal, the following methods are necessary: grinding, wet rendering (which involves the use of steam under pressure), and drying. Special steamed bone meal is a specific type of bone meal obtained by cooking dried bones with steam under pressure, after the removal of grease and meat fibers.

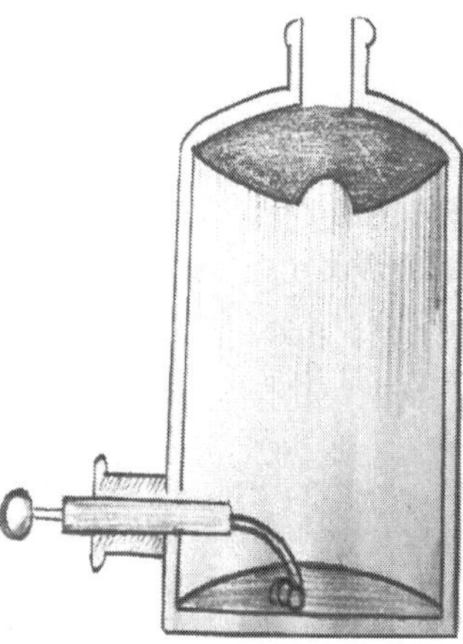

Figure 9: Bone digester

4. **Tankage:** Tankage, also referred to as meat-meal tankage, digester tankage, or feeding tankage, is a high-protein material derived from animal tissues through a method known as wet rendering, which utilizes direct steam. This dried and finely milled substance exhibits variations in its precise composition based on the raw materials and techniques employed. It is commonly recognized by various terminologies depending on its preparation, such as tankage, meat meal tankage, or digester tankage. Generally, it comprises 55-60% protein. Initially, tankage was the solid byproduct obtained after the extraction of animal fat and was utilized as fertilizer due to its richness in nitrogen, phosphorus, and calcium. Over time, its substantial protein, calcium, phosphorus, and fat content rendered it a significant component of animal feed. If tankage contains more than 4.4% phosphorus, it is specifically categorized as digester tankage, which may encompass various forms such as bone meal or meat and bone meal tankage.
5. **Meat meal or meat scraps:** Meat meal or meat scraps are produced from meat remnants processed through dry rendering in steam-jacketed tanks. This technique, which involves the application of steam to heat the material, improves the quality of the protein and typically results in a product with a high protein concentration of approximately 60%. Should the product contain more than 4.4% phosphorus, it is required to be labelled as meat and bone meal or meat and bone scrap. In contrast to tankage, which is generated through wet rendering, meat meal benefits from the regulated dry rendering process, leading to a superior quality protein product.

6. **Blood meal:** Dried and ground blood, referred to as blood meal, is abundant in protein and particularly high in lysine, although it possesses low digestibility. Blood flour is a more refined variant of this product, created from the same dried blood but processed into a finer powder. Both products are utilized in animal feed; however, due to their low digestibility, they are generally employed in restricted amounts.
7. **Feather meal:** Feather meal is produced from feathers that undergo wet rendering and are subsequently ground into a high-protein meal, which can contain as much as 80% protein. Nevertheless, the protein found in feather meal lacks certain essential amino acids, thereby potentially limiting its nutritional efficacy in animal feed.
8. **Poultry by-products meal:** Poultry by-product meal bears resemblance to meat scrap regarding its composition, appearance, and nutritional profile. It is derived from the assorted by-products generated during poultry processing and possesses similar levels of protein, fat, and minerals.

Table 7: Composition of meat cum bone meal

Constituents	**Percentage (%)**
Crude protein	55 (Minimum)
Crude fat	7 (Maximum)
Calcium	10 (Minimum)
Phosphorous	5.0 (Minimum)
Clostridium spores, salmonella and *E. coli*	Nil
Moisture	7 (Maximum)
Pepsin digestibility	85

Basic Information on Stock Feed

Understanding the composition of raw materials is essential for calculating the expected yield of meals derived from various offals. The composition of meats can differ significantly based on the species of animals and the specific parts from which they originate. Regardless of the raw material type utilized, the objective is to achieve a final product with an approximate composition of 85% protein and minerals, 7% fat, and 7% moisture. Excess moisture or fat can diminish the keeping quality of the product, which is undesirable. Since stock feed is not produced from a single type of raw material, the ratio of each ingredient will affect the overall yield. For instance, if only lungs, tripe, and intestines with a very high-water content are used, the yield will be significantly lower. A commercial meal can only be produced without fat removal if the original fat content in the raw materials does not exceed 3.5%, although it is common to encounter higher fat percentages (over 7%) in raw materials. Therefore, to achieve the desired fat content, fat expulsion is necessary. In addition to the intrinsic properties of the raw materials, the yield of the final product is influenced by several other factors.

a) Type of rendering method used i.e., wet rendering (soluble and suspended materials is lost) or dry rendering (all the components are fully recovered).
b) Speed of processing.
c) Type of equipments used.
d) Experience and skill of the operators, which contribute to the gain or loss in the final yield.

In general, the conversion ratio of raw materials to dry meal is:

- 3:1 in dry rendering process
- 4:1 in wet rendering process
- 5:1 in blood meal preparation

Commercial Terms Used for Various Products

Commercial markets accept a definite terminology for animal by-products. They include:

1. **Raw bone meal:** The term "raw bone meal" can be misleading. It actually refers to meal made from bones that have been cooked in an open kettle without steam pressure. This process sterilizes the bones, making them safe for use. Raw bone meal typically has a high protein content, generally no less than 23%, since only a small amount of ossein (collagen) is removed. The average composition of raw bone meal includes:

Protein	:	26%
Calcium	:	23%
Phosphorus	:	11%
Fat	:	Acceptable level

   The exact composition can vary depending on the specific processing methods and source of the bones.

2. **Steamed bone meal:** Bone meal produced via steam pressure, regardless of whether it is through dry or wet rendering, leads to the extraction of a significant portion of proteins and fats. The typical composition of this type of bone meal generally comprises:

Protein : 7%

Calcium : 32.5%

Phosphorus : 15%

This process produces a mineral-rich meal suitable for animal feed.

3. **Bone ash:** Bone ash is created through the combustion of bones in an open environment, which permits unrestricted oxygen access. It possesses a substantial phosphorus concentration, generally between 15.3% and 16.6%. The combustion process eliminates organic substances, resulting in a residue that is predominantly mineral-based, with phosphorus being a key element.
4. **Meat meal or Digester tankage or Meat scrape or Dry rendered tankage:** This is derived from the wet or dry rendering of animal tissues and must adhere to specific standards; it should not exceed 10% phosphoric acid (which is equivalent to 4.4% phosphorus) and must contain a minimum of 55% protein. Materials such as hooves, horns, hair, blood, manure, and stomachs or intestines are prohibited from being utilized in this process.
5. **Meat and bone scrape or Meat and bone meal or Carcass meal:** It refers to a product characterized by a protein content of less than 55% and a phosphorus content exceeding 4.5%. Bones constitute a significant portion of the raw materials used for this product.
6. **Blood meal:** Blood meal is notable for its exceptionally high protein content, frequently surpassing 80%, while being low in calcium and phosphorus. High-quality blood meal should contain approximately 85% protein and be devoid of fat, fiber and phosphate of lime.
7. **Liver meal:** Liver meal is produced by drying and grinding the livers of slaughtered animals, often utilizing condemned livers for this purpose. To comply with quality standards, liver meal must contain a minimum of 27 milligrams of riboflavin per pound.

The market price for the majority of these products, with the exception of bone meal, is influenced by their protein content and is calculated based on "Protein units," which denote each percent of protein found in a ton of the meal. In contrast, the pricing of bone meal is determined by its phosphorus content.

Rendering of Animal Fat

Fats derived from animal carcasses, which fluctuate based on water content, constitute 18-30% of the carcass weight in market steers and 12-20% of the live weight in average market hogs. These fats, sourced from slaughtered livestock and poultry, can be classified as either edible or inedible. The primary sources include beef and pork, with lesser amounts obtained from sheep and goats. Animal fats are mainly categorized into:

a) **Killing fats:** These fats are removed from animals during the slaughter process.

b) **Cutting fats:** These fats are trimmed from the carcasses once they are removed from the chiller.

Both killing fats and cutting fats can be utilized to produce lard and tallow for consumption. Conversely, inedible fats are derived from deceased or diseased animals and non-prime parts of carcasses. These fats are typically extracted by cooking or by subjecting fatty tissues to low temperatures, which facilitates the release of fat primarily through the physical rupture of fat cells.

1. **Continuous low temperature dry rendering:** Continuous low-temperature dry rendering is a method for producing high-quality fats through a continuous process of heating, separating, and cooling. The process involves following steps are:

   a) Mincing of raw material.

   b) Melting by live steam injection at 90°C.

   c) Continuous separation of solids from the liquid fat in a decanter centrifuge.

   d) Further heating.

   e) Centrifugation.

   f) Cooling in a plate heat exchanger below solidification point.

This approach prevents the breakdown of fat by lipase, which functions optimally at temperatures ranging from 40°C to 60°C but becomes inactive at temperatures exceeding 60°C. Consequently, it reduces chemical activity, burning, oxidation, and the development of undesirable flavours. In terms of space and labour efficiency, it surpasses dry batch rendering; however, it does not facilitate the hydrolysis of hair and wool due to the absence of pressure cooking.

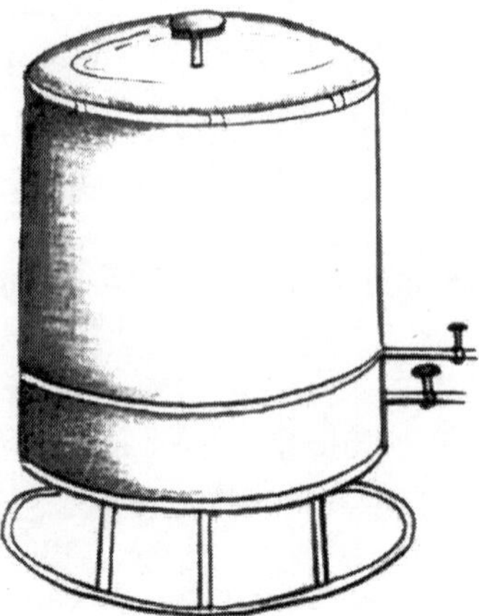

Figure 10: Fat extractor

2. **Steam rendering process:** The steam rendering technique is frequently employed for adipose tissue, wherein the substance is situated in a sizable tank featuring a conical base. Live steam is introduced into the tank, and rendering takes place under pressure to expedite the cooking process. Upon completion of the procedure, the pressure is slowly alleviated, permitting the mixture to settle prior to the extraction of the adhered water.

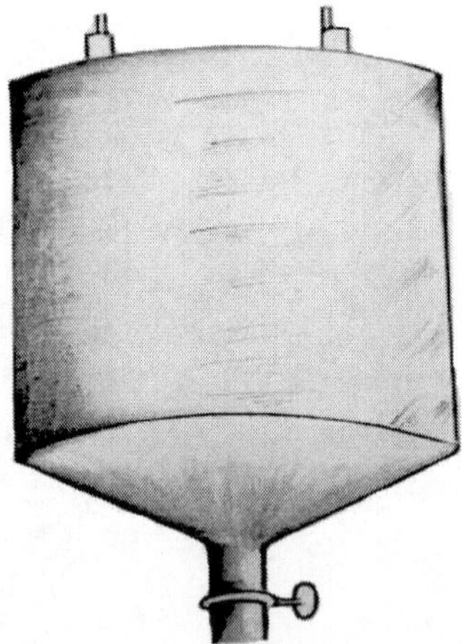

Figure 11: Steam renderer for fat extraction

3. **Dry rendering of fat:** Inedible tallow and grease are mainly generated via the dry rendering technique. In this approach, fatty tissue is introduced into a horizontal steam-jacketed cylinder equipped with rotating internal

blades that come close to the walls. The fat is subsequently cooked under high pressure, atmospheric pressure, or in a vacuum. Following the cooking process, the material is filtered to eliminate cracklings, and the extracted fat is gathered as technical fat. Other methods of dry rendering consist of open kettle, drip, and caustic rendering techniques.

Tallow and Lard

Tallow is defined as the rendered fat derived from cattle and sheep, whereas lard is the fat obtained from pigs. The differentiation between tallow and lard is determined by their solidification temperature, referred to as the titre: fats with a titre exceeding 40°C are categorized as tallow, while those with a titre below this threshold are classified as lard. The titre serves as an indicator of the fat's softness or hardness. Inedible tallow and grease find application in livestock and poultry feed as high-energy supplements. These additives contribute to reducing dust, enhancing colour and texture, and improving the palatability of the feed. Furthermore, fatty acids extracted from tallow are employed in a variety of industrial uses, including abrasives, lubricants, shaving cream, candles, cement additives, cleaners, cosmetics, paints, polishes, detergents, soap, plastics, printing inks, and perfumes. Edible tallow and lard are incorporated into products such as oleomargarine and cooking fats.

Use of tallow/lard

a) **Inedible tallow:** Inedible tallow, derived from animal fat, serves as a high-energy additive in livestock and poultry feed. It is stabilized with antioxidants to prevent rancidity and provides several benefits in animal feed:

- Reduces dustiness
- Improves colour and texture
- Enhances palatability
- Increases pelleting efficiency
- Minimizes wear and tear on machinery

Fatty acids from inedible tallow are used in various products, including:

- Shaving cream
- Asphalt tiles
- Candles
- Caulking compounds
- Cement additives
- Cleaners

- Cosmetics
- Deodorants
- Paints
- Polishes
- Perfumes
- Detergents
- Plastics
- Printing ink
- Synthetic rubber
- Water-repellent compounds
- Soap making.

b) **Edible tallow:** Edible tallow and lard are utilized in

- Oleomargarine (margarine)
- Shortenings
- Cooking fats

Tallow is often preferred for its flavour in fried foods compared to vegetable oils.

4

Utilization of Blood

Blood serves as a significant byproduct within the meat industry, and the efficient management of its collection and processing is crucial for improving profitability. The amount of blood obtained from slaughtered animals typically ranges from 5-7% of their live weight. Despite diligent efforts, a portion of blood is inevitably lost during the slaughtering process due to spillage and various other factors. To reduce these losses, several essential practices can be adopted:

1. **Effective slaughterhouse design:** Enhancing the layout and operational procedures within the slaughterhouse can significantly minimize spillage. This encompasses the implementation of well-structured drainage systems and collection vessels.
2. **Training and protocols:** It is essential to ensure that personnel are thoroughly trained in blood collection methods and strictly follow established protocols to mitigate accidental loss. This involves appropriate handling and prompt processing to prevent clotting and spoilage.
3. **Technology:** Allocating resources towards advanced blood collection and processing technologies can enhance operational efficiency. Automated systems facilitate the collection, filtration, and storage of blood while minimizing waste.
4. **Regular maintenance:** Maintaining equipment and facilities in optimal condition is vital to prevent leaks and other complications that may result in blood loss.
5. **Storage and processing:** After collection, blood must be processed promptly to prevent coagulation. Efficient cooling and storage practices are essential for preserving its quality and maximizing its potential for subsequent processing into products such as blood meal or plasma.

Collection of Blood

Blood may be collected directly into metal or plastic drums when animals are suspended for bleeding. In cases where the animals are slaughtered on the ground, small enamel or plastic bowls can be placed under the bleeding site to collect the blood, which can subsequently be transferred into a drum. The anticipated yield of blood from average tropical livestock is generally as follows:

Table 8: Expected yield of blood from tropical livestock

Species	Yield of blood
Buffalo or cattle	10-12 kg
Sheep or goat	1-1.5 kg
Pig	2-3 kg
Poultry	30-50 g

Most of the animal blood require proper disposal from the slaughter house for following reasons:

1. Blood will clot within few minutes and choke the drainage.
2. Decomposition of blood will affect the sanitation and spread foul smell in the premises.
3. It will attract flies, crows, rodents etc.

Utilization of Blood

Animal blood serves multiple purposes, and the method of its collection is depending upon its intended use. Below are some applications of blood:

1. **As a food source for humans:** Blood derived from slaughtered animals is a significant by-product within the meat industry, presenting opportunities as a source of diverse food ingredients, especially in meat products. It can be converted into specialized nutritional and functional components such as emulsifiers, texturizers, colorants, and bioactive substances. The industry encounters the challenge of enhancing bleeding and recovery techniques to render this by-product economically feasible. While not extensively utilized as food in numerous cultures today, blood holds historical and cultural significance in various societies.
   a) **Blood sausages and black pudding:** In Europe, animal blood is conventionally employed to produce blood sausages, blood pudding, biscuits, and bread. For instance, black pudding in the United Kingdom is a type of blood sausage created by combining blood with fat, grains, and spices. In Asia, blood is incorporated into dishes such as blood curd, blood cake, and blood pudding.
   b) **Blood soups:** In certain cultures, traditional soups or stews prominently feature blood as a key ingredient. These meals are generally rich and hearty, utilizing blood for its unique flavour and nutritional advantages.

c) **Blood as a nutrient:** Blood is abundant in vital nutrients like iron and protein, which can be particularly advantageous in diets that lack sufficient meat or for individuals with specific dietary requirements.

d) **Fibrin:** Fibrin, a fibrous protein essential for the clotting of blood, can be extracted and used in food products as a binding agent or protein source. In the realm of food science, it is esteemed for its functional characteristics, such as gel formation and emulsion stabilization.

e) **Peptone:** Peptone, which is obtained from the breakdown of proteins, is mainly utilized as a growth medium in the fields of microbiology and biotechnology. While it is not generally employed directly in food intended for human consumption, it has an indirect influence on food production, particularly in fermentation processes for additives or probiotics. Its application is highly specialized and regulated to ensure safety and compliance with food standards.

f) **Ritual and symbolism:** In numerous cultures, the act of consuming blood can carry ritualistic or symbolic meaning, often embodying ideas such as strength, vitality, or a connection to the spiritual realm.

g) **Modern uses:** Although less prevalent in contemporary cuisine, blood products, including blood pudding and other dishes made from blood, continue to be favoured in areas with a rich culinary heritage. Furthermore, blood plasma is utilized as a protein enhancer in various foods and acts as a binder or stabilizer in meat products.

The incorporation of blood in food is regulated to mitigate health risks, such as ensuring adequate cooking to eliminate pathogens. Attitudes towards blood consumption vary significantly across cultures, with some individuals viewing it as a delicacy while others opt for more conventional food sources.

2. **In industrial use:** Blood, which is abundant in proteins and other valuable elements, has numerous industrial uses that extend beyond conventional food applications. Some significant applications include:

a) **Biomedical products:** Blood serves as a source of plasma proteins such as albumin and immunoglobulins, which are essential in the pharmaceutical sector for the production of drugs, vaccines, and diagnostic reagents. These proteins are isolated from blood plasma through methods like fractionation.

b) **Industrial adhesives:** Blood proteins, particularly fibrinogen, are incorporated into adhesive formulations. Fibrin glue, derived from blood plasma, is employed in surgical operations to bond tissues and facilitate wound healing.

c) **Animal feed:** Blood meal, a byproduct from slaughterhouses, is a high-protein additive utilized in animal feed. It supplies vital amino acids and nutrients for livestock and aquaculture, promoting healthy growth and muscle development due to its high protein content and bioavailability.

d) **Biodegradable plastics:** Blood proteins are being explored as a sustainable source for biodegradable plastics. When altered and processed, these proteins can be utilized to produce materials suitable for packaging and biomedical uses.

e) **Fertilizers:** Blood meal is applied in agriculture as an organic fertilizer due to its high nitrogen and nutrient content. It improves soil fertility and encourages plant growth.

f) **Industrial chemicals:** Products derived from blood can be processed to extract chemicals such as haemoglobin-based oxygen carriers for biomedical applications and iron-based compounds for industrial purposes.

g) **Biotechnology:** Components of blood, including enzymes and growth factors, are employed in biotechnology for enzyme production and as supplements in cell culture media for bio-pharmaceutical manufacturing.

h) **Plywood industry:** In the plywood sector, blood is sometimes utilized as a binder or waterproof adhesive. The proteins present in blood can act as a natural adhesive when mixed with other additives, improving the adhesion of wood veneers in plywood manufacturing.

i) **Tanneries:** In the process of leather tanning, blood is incorporated into tanning solutions to stabilize collagen fibers. The salts present in blood enhance the tanning procedure, thereby improving the quality of leather and imparting a light-coloured protein finish along with a glossy appearance.

j) **Textile dyeing:** Blood has been studied as a natural mordant in the dyeing of textiles. Mordants are agents that assist in fixing dyes to fabrics, and the components found in blood may enhance dye adherence or modify the properties of fabrics during the dyeing process.

k) **Fire extinguishers:** Blood proteins, particularly albumin, have been researched for their potential application in fire extinguisher formulations due to their heat-resistant characteristics. Albumin can form a protective layer on surfaces, which aids in preventing oxygen from exacerbating fires.

l) **Ceramics:** Proteins derived from blood can serve as additives in the production of ceramics, acting as binders or glazing agents. These proteins contribute to the strength and finish of ceramic products.

m) **Cosmetic industry:** Blood plasma and serum proteins, such as albumin and globulins, are employed in cosmetics and skincare formulations. They act as moisturizers, film formers, or texturizers in a variety of products.

n) **Lithography:** Historically, blood albumin has been utilized in the preparation of inks for lithographic printing. The presence of albumin stabilizes the ink and enhances its viscosity, facilitating the effective transfer of images onto paper or other surfaces.

o) **Substitute for egg albumen:** Blood albumin has been explored as a substitute for egg albumen in numerous applications. It can fulfil similar roles, such as stabilizing foams, emulsifying ingredients, and binding components in food products. This alternative is particularly beneficial for individuals with egg allergies or specific dietary restrictions.

Table 9: Uses of animal blood in various industries

Food	Emulsifier, stabilizer, clarifier, colour additive, nutritional component.
Feed	Lysine supplement, vitamin stabilizer, milk substitute, nutritional component.
Fertilizer	Seed coating, soil pH stabilizer, mineral component.
Laboratory	Tissue culture media, haemin, blood agar, peptone, glycerophosphates, albumin, globulin, sphingomyelin, catalase.
Medicine	Agglutinin test, immunoglobulin fractionation techniques, blood clotting factors, sutures, fibrinogen, fibrinolysin, fibrin products, serotonin, plasminogen, plasma extenders.
Other industry	Adhesive, resin extender, finishes for leather and textiles, insecticide spray adjuvant, egg albumin substitute, foam fire extinguisher, porous concrete, ceramic and plastic manufacturer, plastic and cosmetic base formulations.

These industrial applications demonstrate the versatility of blood as a raw material beyond its traditional use in food. However, they require careful processing and strict adherence to regulatory standards to ensure safety and effectiveness in each field.

3. **As livestock feed:** Blood, especially in the form of blood meal, serves as a significant element in livestock nutrition due to its rich nutrient profile. It enhances health, growth, and productivity in various animal sectors, including poultry, swine, and aquaculture. Below are some essential applications and advantages of blood as livestock feed:

a) **Protein source:** Blood meal is a high-protein additive derived from blood gathered at slaughterhouses. It is abundant in essential amino acids vital for animal growth, muscle development, and overall well-being. With a protein concentration generally between 80% and 90%, it acts as a concentrated dietary protein source for livestock.

b) **Nutrient content:** Beyond proteins, blood meal supplies vital nutrients such as iron, phosphorus, and trace minerals like zinc and copper. These nutrients are essential for metabolism, bone growth, and immune system functionality in animals.

c) **Palatability and digestibility:** Livestock usually find blood meal appealing, which encourages consumption. Its high digestibility guarantees effective nutrient absorption by animals.

d) **Feed efficiency:** The inclusion of blood meal in livestock diets can improve feed efficiency and overall performance by optimizing the balance of amino acids and other necessary nutrients for growth and production.

e) **Cost-effective:** Blood meal represents a cost-efficient protein source when compared to traditional feed ingredients. It provides a sustainable approach to utilizing slaughterhouse byproducts, minimizing waste and promoting economic sustainability in livestock farming.

The incorporation of blood and blood products in livestock feed is subject to regulations to ensure safety and quality. This encompasses appropriate processing to eliminate contaminants and preserve nutritional value.

4. **Fertilizer:** Blood meal, a byproduct of animal blood collected at slaughterhouses, is utilized as an organic fertilizer in both agriculture and horticulture. Key features of its application include:

a) **Nutrient composition:** Blood meal is rich in nitrogen, an essential nutrient for the growth and development of plants, typically comprising approximately 12-15% nitrogen. This positions it as a potent nitrogen source for vegetation.

b) **Gradual nitrogen release:** In contrast to synthetic nitrogen fertilizers that release nitrogen rapidly, blood meal offers a slow and consistent nitrogen release over time. This gradual process guarantees a reliable nitrogen supply to plants while minimizing the risk of nitrogen leaching into groundwater.

c) **Organic constituents:** Beyond nitrogen, blood meal is composed of organic matter and trace elements of other nutrients such as phosphorus and potassium. These components contribute to enhanced soil fertility and promote overall plant vitality.

d) **Soil enhancement:** Blood meal contributes to the improvement of soil structure and texture by stimulating microbial activity and boosting soil fertility. Additionally, it aids in maintaining soil pH balance and increasing nutrient accessibility for plants.

e) **Seed treatment:** In agricultural practices, seed coatings are employed to improve germination rates, protect against pests and diseases, and supply nutrients. Blood is also utilized in seed coatings for these functions.

f) **Organic agriculture:** The application of blood meal as a fertilizer aligns with organic farming and sustainable agricultural practices by decreasing dependence on synthetic fertilizers, reducing environmental impact, and promoting soil health and biodiversity. Blood meal is typically applied to the soil surrounding plants or incorporated into the soil prior to planting. It is especially beneficial for crops with high nitrogen requirements, such as leafy greens, vegetables, and fruits.

The utilization of blood meal as a fertilizer is subject to regulations to ensure safety and quality. Proper processing and handling are essential to maintain its nutrient integrity and effectiveness.

5. **In biochemical and pharmaceutical fields:** Blood and its derivatives are employed in a variety of biochemical and pharmaceutical applications due to their distinctive properties and biological functions. Common applications include:

   a) **Albumin synthesis:** Albumin, a protein derived from blood plasma, serves as a stabilizing agent in pharmaceuticals for drugs and vaccines. It plays a crucial role in maintaining osmotic pressure and ensuring the stability of formulations.

   b) **Clotting factors VIII and IX:** These factors, obtained from blood plasma, are utilized in the treatment of haemophilia, a hereditary condition characterized by improper blood clotting.

   c) **Immunoglobulins:** These antibodies, sourced from blood plasma, can be purified for therapeutic use to enhance the immune response or address specific immune-related disorders.

   d) **Heparin:** Heparin, an anticoagulant derived from animal blood, particularly from pigs and cows, is employed in clinical settings to avert blood clot formation during surgical procedures and to manage thrombotic conditions.

e) **Serum:** Blood serum, the fluid component of blood following the removal of clotting factors, is utilized in diagnostic assays and research due to its rich content of proteins, electrolytes, and antibodies.

f) **Haemoglobin:** Haemoglobin, the protein responsible for oxygen transport in red blood cells, is applied in medical therapies as an oxygen carrier in blood substitutes.

g) **Cell culture media:** Components from blood, such as foetal bovine serum (FBS), are incorporated into cell culture media to provide vital nutrients and growth factors necessary for cell proliferation and maintenance in laboratory settings.

h) **Serotonin:** Serotonin, derived from blood, is utilized in pharmaceuticals to address conditions like depression, anxiety disorders, and certain neurological issues. Selective Serotonin Reuptake Inhibitors (SSRIs) represent a prevalent category of medications that alter serotonin levels in the brain.

i) **Peptone:** Peptones, obtained from blood, are nutrient-dense substances formed from partially digested proteins. They are employed in microbiological applications to cultivate bacteria, fungi, and other microorganisms in laboratory environments, supplying essential nutrients for their growth and reproduction.

j) **Plasma extenders:** These solutions are specifically formulated to either replace or augment blood plasma in various medical treatments. They replicate the osmotic and volumetric characteristics of blood plasma and are employed in emergency medicine and surgical procedures to restore blood volume and sustain blood pressure when whole blood or its components are either unavailable or inappropriate.

k) **Thrombin:** This enzyme, present in blood plasma, plays a vital role in the coagulation process by converting fibrinogen into fibrin, thereby facilitating clot formation. In medical practice, thrombin is utilized to enhance clotting during surgical operations, manage haemorrhaging, and address specific types of wounds. Additionally, thrombin is instrumental in the production of fibrin glue and sealants for surgical use.

l) **Prothrombin:** As a protein precursor to thrombin, prothrombin is synthesized in the liver and is transformed into thrombin with the assistance of calcium ions and Factor Xa during the coagulation cascade. Prothrombin Complex Concentrates (PCCs), which comprise prothrombin along with other clotting factors, are employed in the treatment of bleeding disorders such as haemophilia and vitamin K deficiency.

6. **In laboratory and bacteriological media:** Blood serves as a specialized growth medium in laboratory and bacteriological environments for certain organisms. The utilization of blood is as follows:

   a) **Enriched media:** Blood agar is a widely utilized medium that incorporates blood, typically sourced from sheep or horses, into a basic agar formulation. This addition enriches the medium with vital growth factors, including vitamins (notably B vitamins), amino acids, and other essential nutrients required by fastidious organisms (those with intricate growth needs). Blood agar is employed in clinical microbiology for the isolation and identification of bacteria from samples such as throat swabs, wound swabs, and blood cultures. It is particularly effective in identifying pathogens responsible for infections, including organisms like *Haemophilus influenzae*, which necessitate specific factors found in blood for their growth. The composition of blood agar can be modified to include specific concentrations of these factors to enhance the accuracy of testing.

   b) **Nutrient source:** Blood serves as a natural reservoir of nutrients, including iron, which is vital for the proliferation of numerous bacteria.

   c) **Anaerobic growth:** Blood can be utilized under anaerobic conditions to facilitate the growth of anaerobic bacteria, providing a more authentic environment in comparison to synthetic media.

   d) **Tissue culture media:** Components of blood, such as serum, which contains albumin, globulins, and various other proteins, play a significant role in tissue culture media. These media are formulated to mimic the natural conditions necessary for cell growth and proliferation. Albumin supplies essential proteins and amino acids, while globulins aid in immune modulation and other cellular activities. Tissue culture media are extensively employed in cell biology research, vaccine development, and medical diagnostics.

   e) **Sphingomyelin:** Sphingomyelin, a sphingolipid that is essential for cellular membranes, particularly in neuronal cells, is generally not directly incorporated into bacteriological media. Nevertheless, investigating sphingomyelin metabolism can be crucial for comprehending bacterial interactions with host cells or for research in lipid metabolism.

f) **Catalase:** Catalase is an enzyme that catalyzes the decomposition of hydrogen peroxide into water and oxygen. In bacteriological media, it is utilized in biochemical assays to distinguish bacteria based on their capacity to produce catalase. This assay is instrumental in differentiating staphylococci (catalase-positive) from streptococci (catalase-negative).

Blood Products

1. **Plasma, serum and blood albumen:** Blood is composed of plasma, which serves to suspend erythrocytes, leucocytes, and thrombocytes (platelets). Plasma is categorized into serum and fibrinogen. The enzyme thrombin acts on fibrinogen to generate fibrin. Plasma is isolated from un-clotted blood through centrifugation, whereas serum is derived from clotted blood. The typical clotting duration for the majority of domestic animals ranges from 3 to 6 minutes. Blood intended for human use is exclusively collected in abattoirs that are equipped with bleeding rails. A trocar knife, which features a hollow handle, along with a canula, is employed for the hygienic collection of blood, which is subsequently stored in clean, sterile stainless-steel containers. The optimal container should measure 45 cm in diameter and 15 cm in depth, and it can be sanitized using hypochlorite or through steam sterilization.
   - **Plasma production:** Plasma represents the liquid fraction of blood that remains after the removal of cells (including red blood cells, white blood cells, and platelets). It is composed of approximately 90% water and contains proteins, electrolytes, hormones, gases, and waste products. Plasma is obtained by centrifuging whole blood that has been collected in anticoagulant tubes, such as Trisodium citrate, Oxalate, EDTA, or Heparin. This procedure results in the separation of plasma at the top, a buffy coat (which contains white blood cells and platelets) in the middle, and red blood cells at the bottom. Plasma is then preserved in bottles or polyethylene bags and is frozen for therapeutic applications, including plasma transfusions and the production of plasma-derived medicinal products.
   - **Serum production:** Serum constitutes the liquid component of blood that remains after the removal of clotting factors (such as fibrinogen) and blood cells from clotted blood. It bears resemblance to plasma but is devoid of clotting factors due to their utilization during the clotting process. The preparation of serum involves the collection of blood, allowing it to clot, and subsequently processing it in a cold environment. The chilled clotted blood is then segmented into

smaller fragments to expedite the contraction of the clot. Serum that is harvested within the initial 12 hours is typically clear, containing minimal suspended red blood cells (RBCs). Following this, it undergoes centrifugation at 1000 rpm and is filtered using a Seitz filter. The resultant yield is approximately 10-12% of the total weight of the whole blood. Serum is preserved at temperatures of 4-5°C for a duration of up to one month and at -20°C for six months. It is utilized in diagnostic assays, research, and in vitro studies where the presence of clotting factors may pose interference. Both plasma and serum are maintained at low temperatures, generally ranging from -20°C to -80°C, to safeguard their biochemical characteristics and to avert the degradation of proteins and other constituents. To ensure the integrity of these samples for precise laboratory analysis and research, it is crucial to handle them appropriately and to prevent repeated freeze-thaw cycles.

- **Blood albumen:** Dried blood serum, referred to as blood albumen, serves as a cost-effective alternative to dried egg albumen powder. It is produced by incorporating 0.05% phenol into clear yellow serum, followed by spray or vacuum drying to create a fine powder. Blood albumen yields between 10-20% of the serum's weight and remains stable when stored in airtight containers in cool environments for several months.

2. **Fibrinogen:** Blood that has been collected using anticoagulants such as citrate, EDTA (Ethylene Diamine Tetra Acetic acid), or heparin undergoes centrifugation to isolate plasma. This plasma is subsequently buffered with a sodium citrate-acetic acid solution while aqueous ethanol is added until the mixture achieves a pH of 7.2 and an ethanol concentration of 8%. This procedure results in the precipitation of fibrinogen, which is then extracted through centrifugation. The plasma is then filtered, sterilized, freeze-dried, and stored.

3. **Fibrin foam, powder, and bioplasts:** Fibrin foam is a porous substance that ranges in colour from white to cream and possesses a remarkable capacity to absorb liquid, with the ability to take in up to 30 times its weight. It is especially beneficial in surgeries involving nerves and arteries, as well as in prostate procedures. Fibrin foam does not hinder the efficacy of antibiotics and is ultimately absorbed by the body. The method for preparing fibrin foam from animal blood adheres to principles akin to those applied to human blood, with modifications tailored to the specific animal species and intended use:

a) **Blood collection:** Blood is drawn from the chosen animal species into sterile containers containing an anticoagulant to avert clotting. The selection of anticoagulant may differ based on the species and the specific application of the fibrin foam.

b) **Separation:** The collected blood is subjected to centrifugation to divide it into its components: red blood cells, white blood cells, platelets, and plasma. Additional centrifugation is performed to obtain platelet-poor plasma (PPP), which is essential for optimal fibrin formation.

c) **Fibrin formation:** In platelet-poor plasma (PPP), fibrinogen undergoes activation to yield fibrin. This process is generally facilitated by the addition of thrombin or calcium chloride ($CaCl_2$). Thrombin facilitates the conversion of fibrinogen into fibrin monomers, which subsequently polymerize to form fibrin strands.

d) **Foam formation:** The solution of fibrinogen, now transformed into fibrin, is subjected to agitation or whipping to incorporate air and generate bubbles within the fibrin matrix. Additionally, mechanical techniques or gas-producing substances may be employed to create foam. Crosslinking agents such as $CaCl_2$ can be introduced to enhance the stability of the fibrin network.

e) **Shaping and setting:** The fibrin foam is molded as required and permitted to set. During this phase, the polymerization of fibrin fortifies the structure of the foam.

f) **Quality control and sterilization:** The fibrin foam is subjected to stringent quality control measures to verify sterility, biocompatibility, and mechanical integrity. Sterilization techniques, including gamma irradiation or ethylene oxide gas, are utilized to ensure the product's safety. Animal-derived fibrin foam finds application in veterinary medicine for wound healing, tissue engineering, and as a haemostatic agent in surgical interventions. It also holds significance in research for investigating tissue regeneration and drug delivery.

- **Fibrin powder:** Fibrin powder is employed to manage bleeding in scenarios where coagulation is impaired, such as in skin injuries. It originates from a fibrinogen solution that develops a gel-like consistency upon polymerization under regulated conditions. This gel is subsequently freeze-dried or spray-dried to yield fibrin powder, which is further refined to achieve the desired particle size and packaged in a sterile manner to maintain its integrity. In surgical contexts, fibrin powder serves as a haemostatic agent, a scaffold for tissue engineering, and in regenerative medicine. It is also instrumental in research for examining cellular behaviour, wound healing, and drug delivery owing to its biocompatibility and regenerative characteristics.

- **Fibrin bioplasts:** These are materials shaped to resemble plastic, created from fibrin powder. They can be formed into various configurations, including bone joints, and do not require post-healing removal from the body.

4. **Blood meal:** In slaughterhouses, blood is frequently either disposed of in the sewage system or processed alongside other waste to manufacture products such as animal feed and technical materials. Blood meal, a dark brown, dehydrated substance with a moisture content of 8-12%, is produced by drying whole blood. The yield of blood meal is approximately 20% of the total blood, and it is abundant in protein (80-85%) and essential amino acids, including lysine (6-9%), tryptophan, and methionine. Blood meal serves as both animal feed and fertilizer (comprising 12% nitrogen, 0.22% phosphorus, and trace elements). It is occasionally combined with super phosphate to create compound fertilizers and can also function as an adhesive in asphalt emulsions, insecticides, and ceramics. The preparation of stock feed is an efficient method for managing slaughterhouse waste while generating considerable revenue. This process includes the following steps:

a) **Blood collection:** Blood must be collected meticulously to prevent contamination from floor washings, detergents, insecticides, or other foreign substances. Ideally, blood should be processed on the same day it is collected. It is collected from slaughtered animals into sterile containers or tanks, which are equipped with refrigeration to inhibit bacterial growth and spoilage. Maintaining controlled conditions, typically below 4°C, is crucial to keep the blood fresh and prevent coagulation.

b) **Storage and transport:** Blood is transported to blood meal production facilities under refrigerated conditions to preserve its quality and safety. Alternatively, whole blood can be stored and transported using one of the following methods:

i. Combine whole blood with an equal volume of rice bran, which effectively absorbs the liquid. This blend can be transported in hessian sacks during wet or cool seasons and dried in the summer to achieve a moisture level of 10-12%.

ii. Introduce 1% quick lime by weight into the blood, resulting in a black, rubber-like texture. This addition prevents sticking to containers, extends storage duration for 24 hours, and diminishes fly attraction. Furthermore, it enhances the calcium content of the final product.

iii. Incorporate 20% common salt by weight into the blood to prolong its shelf life and ensure safe transport to the production facility.

c) **Heat treatment:** The blood mass is subjected to cooking to sterilize and decrease its moisture content, typically utilizing one of the following techniques:

i. **Direct fire cooking:** The blood mass is continuously stirred and cooked directly over an open flame. This method requires maintaining a steady temperature and stirring to guarantee even cooking and complete sterilization.

ii. **Double jacketed boiler:** Alternatively, the blood mass can be cooked in a double jacketed boiler, where steam circulates through the outer jacket to heat the mass. The standard procedure involves heating to 100°C for approximately 30 minutes. This technique ensures uniform heating and aids in effective sterilization while minimizing moisture content.

d) **Pressing:** Following cooking, the blood mass is placed in hessian or other porous bags and suspended to drain. This procedure eliminates 40-45% of the moisture. Employing a screw press can significantly accelerate this drying process by effectively extracting moisture, thereby reducing overall drying time.

e) **Drying:** To diminish moisture content and stabilize the protein, the pressed blood solids can be dried using various methods:

i. **Sun drying:** In hot climates, the blood mass is spread out in shallow trays and sun-dried for 2-3 days.

ii. **Cabinet drying:** In pilot plants, a cabinet dryer equipped with steam coils and an exhaust fan efficiently dries the blood coagulum within 8-12 hours.

iii. **Spray drying:** Blood is atomized into a heated chamber where it rapidly dries into fine particles.

iv. **Drum drying:** Blood is applied to a heated drum, where it dries into flakes that are subsequently ground into powder.

v. **Flash drying:** The blood undergoes brief exposure to high temperatures, which rapidly evaporates moisture while retaining protein content.

f) **Cooling:** Following the drying process, the blood meal is cooled to room temperature to avert condensation and ensure its stability prior to further processing.

g) **Milling:** The dried blood meal is milled to achieve a uniform particle size, which guarantees consistency and enhances handling properties for various applications.

h) **Fumigation:** Fumigation may be performed to eradicate any remaining microbes or pests, utilizing agents such as ethylene oxide, methyl bromide, or phosphine gas. This process is conducted under controlled conditions to ensure the safety of the product and compliance with regulations.

i) **Packaging:** The processed and fumigated blood meal is packaged in sterile polyethylene bags or airtight containers to preserve its quality. The choice of packaging materials is critical to prevent moisture ingress and protect against environmental contaminants.

j) **Storage:** Packaged blood meal should be stored in a cool, dry environment to maintain its nutritional quality and inhibit microbial growth. Effective control of temperature and humidity is essential for prolonging its shelf life. While blood meal typically lasts no more than one month, the addition of lime (CaO) can extend its storage duration to several months. Blood meal is frequently utilized in calf starter rations, as well as in feeds for swine and poultry.

Nutritive value of blood meal: Despite the high protein content of blood meal, it is less digestible compared to meat meal.

Table 10: Nutritive value of blood meal

Particulars	Percent availability
Protein	75 - 85%
Moisture	8 - 12%
Fat	1.2 - 1.6%
Ash	3.8 - 5.6%
Sugar	0.4%
Nitrogen free extract	2.6%

5. **Foam compounds:** To create protein foam compounds derived from blood for application in fire extinguishers, the procedure encompasses several essential stages to transform blood proteins into a foam that is appropriate for fire suppression. Below is a simplified flow diagram illustrating the preparation:

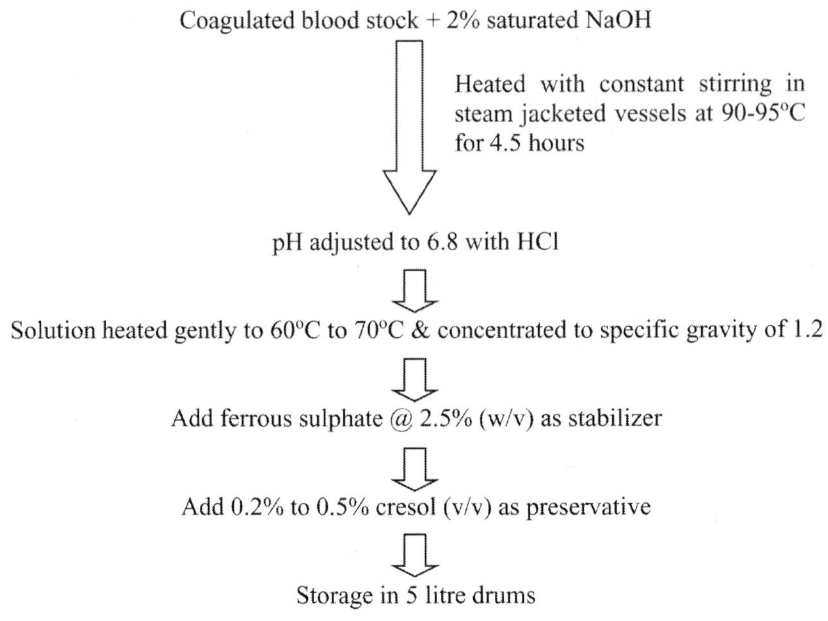

Figure 12: Flow diagram for preparation of foam compound

6. **Organic fertilizer:** Blood that is split and soiled is gathered in drums and transported in sealed vehicles within a timeframe of 4-6 hours for conversion into fertilizers. To ensure the preservation of the blood, it is treated with a 2% formalin or 2% Lysol solution and subsequently dried under sunlight. The dried blood comprises approximately 12% nitrogen, in addition to minor amounts of phosphorus, iron, and copper. It is frequently utilized as a compound fertilizer after being enhanced with super-phosphates. This application is consistent with the principles of organic farming and sustainable agriculture, as it diminishes dependence on synthetic fertilizers, lessens environmental impact, and promotes soil health and biodiversity.

5

Utilization of Bone, Horn and Hoof

Bones, horns, and hooves represent essential by-products of the slaughterhouse sector. Frequently obtained from deceased or fallen animals, these by-products possess considerable applications both independently and in the production of secondary goods. Their use fosters the development of secondary rural industries, consequently creating employment opportunities for individuals in rural regions. Furthermore, by fully utilizing these by-products, producers can secure higher prices for their livestock, while processors may experience enhanced profitability. Effective management of these by-products could also play a role in reducing the costs associated with meat and meat products.

Composition of Bone

Bone constitutes roughly 15% of the weight of a dressed carcass, although this figure may fluctuate due to variables such as breed, age, fat content, and overall health. Generally, the percentage of bone in a dressed carcass is assessed as the weight of the bones in relation to the total weight of the carcass post-dressing. The bone content varies as follows:

- **Young cattle and buffalo:** 15% to 20% of carcass weight.
- **Older cattle and buffalo**: 20% to 25% of carcass weight.
- **Pigs:** 15% to 20% of carcass weight.
- **Sheep and goats:** 25% to 30% of carcass weight.
- **Poultry:** 30% to 35% of carcass weight.

These percentages could increase if the weight accounts for the meat that is still attached to the bones. Furthermore, the marrow located within certain bones, which represents approximately 4-6% of the total carcass weight, is also edible. Typically, fresh bones possess a significant amount of moisture, fat, and protein, with their average composition detailed as follows:

Table 11: Composition of fresh bones

Particulars	Availability
Moisture	50%
Red and yellow marrow	15%
Organic matter	12%
Inorganic matter	23%

Defatted and desiccated bones are made up of organic and inorganic substances in a proportion of 1:2. The principal element of the organic matter is bone collagen, referred to as ossein, which constitutes between 23% and 36% of the organic material. The inorganic component mainly comprises approximately 33% calcium and 15% phosphorus, along with lesser quantities of sodium, potassium, magnesium, and trace elements including copper, zinc, iron, cobalt, and manganese. Red and yellow marrow is primarily composed of fat, accounting for nearly 96% of its makeup. Bones from desert-dwelling animals, which are derived from deceased creatures, are lightweight and mainly consist of dehydrated ossein, calcium, phosphorus, and a reduced amount of minerals.

Classification of Bones

Bones are categorized into two primary types according to their origin:

a) **Green bones:** These originate from freshly slaughtered animals and possess elevated levels of moisture, protein, and fat.

b) **Desert bones:** These have undergone prolonged exposure to bacterial, atmospheric, and insect activity, resulting in the loss of meat, fat, and tendons. They are lightweight and mainly consist of desiccated ossein, calcium, phosphorus, and various other minerals.

For commercial applications, bones are additionally classified into- a) bones utilized as a source of gelatine, and b) bones utilized as a source of phosphate.

Handling of Bones

Bones can be processed using the following methods to prepare valuable byproducts.

a) **Processing under pressure:** The treatment of bones, along with attached meat, tendons, and offals, under steam pressure yields technical grade fat as well as bone and meat meal.

b) **Boiling bones in an open kettle:** The process of boiling bones in an open kettle facilitates the release of adhering materials and dissolves a minor quantity of ossein. Subsequently, the bones are drained, dried, and crushed to create raw bone meal, which may contain as much as 27% protein. This method also recovers fat, which can be classified as either edible quality or technical grade, contingent upon the raw materials utilized.

c) **Boiling bones under pressure:** The boiling of bones under pressure liberates some ossein, after which the bones are dried to produce steamed bone meal. The ossein that is recovered is employed in the preparation of collagen or glue.

d) **Preserving the ossein:** The ossein extracted from steaming bones is preserved for the production of gelatine. Gelatine is produced by subjecting the bones to extended cooking at temperatures not exceeding 70°C to guarantee high-quality gelatine.

e) **Exposure of bones to the environment:** When bones are subjected to environmental exposure for an extended period, they lose their fat content and are termed withered bones. These withered bones retain only ossein and can be utilized either for gelatine production or for conversion into raw bone meal.

Utilization of Bone

In the past, bones served as materials for creating objects such as dice, buttons, and knife handles. In contemporary times, plastics have predominantly taken over these functions. Due to the intricate composition of bones, numerous methods have been established to extract their constituents, including fat, protein, and inorganic substances. Bones can be processed to yield: gelatine and glue, bone meal and fertilizer.

A. **Gelatine and glue:** The production of gelatine represents a highly lucrative application of bones. Gelatine, which is a protein belonging to the albuminoid category, has both consumable and non-consumable (technical) uses. Edible gelatine is derived from fresh bones obtained from animals that have been slaughtered in accordance with stringent hygiene standards. While gelatine and glue share chemical similarities, gelatine is considered a superior product, whereas glue is a lower-quality variant characterized by a darker hue and restricted to non-consumable applications. Pure gelatine is amorphous, transparent, and devoid of colour, flavour, and odour. It becomes brittle when dried, softens when heated, and produces a burnt hair odour upon decomposition. It expands in cold water, absorbing between 5 to 10 times its weight, and dissolves when heated to 30°C. Edible gelatine finds its applications in ice cream, jellies, soft chocolates, capsules, as a binder in tablets, and as a plasma extender in blood transfusions. In commercial contexts, it is utilized as a sizing agent in textiles and leather, as well as in photography. Glue, which is a lower-grade form of gelatine, is employed as an adhesive in plywood, furniture, sandpaper, and gummed tape.

Preparation of crushed bones for gelatine

The process of crushing bones is undertaken to minimize weight and reduce shipping expenses prior to their packaging for manufacturing or export. The preparation of crushed bones for gelatine production is conducted as follows:

1. **Selection of bones:** Choose for long bones such as the femur, tibia, metatarsus, humerus, radius, ulna, and metacarpus.
2. **Cutting:** Employ a motor saw to detach knee caps and knuckles, thereby exposing the marrow to hot water.
3. **Heating or cooking:** Gradually heat the bones in water to a temperature of 85°C for approximately 6 hours. Any fat and other materials that rise to the surface should be skimmed off.
4. **Washing:** After the fat has been removed, wash the bones with warm water.
5. **Drying:** Dry the bones under sunlight on wire netting or, in humid conditions, in a heated room.
6. **Crushing:** Utilize a stone crusher to crush the bones into 1-2 cm cubes. Fragments that pass through a 0.2 cm mesh are classified as bone meal.

Manufacture of gelatine and glue

Gelatine is created through the action of hot or boiling water on collagen or ossein via the hydrolysis process.

$$\text{Collagen} + H_2O \xrightarrow{\text{heat}} \text{Gelatine}$$

The procedure for manufacturing gelatine and glue is outlined below.

1. **Washing:** Bones that have been defatted and uniformly crushed into 1-2 cm cubes are washed with water. Similarly, glue stock undergoes a washing and soaking process. To eliminate non-collagenous materials and fat, both are either sun-exposed or soaked in lime water [a saturated solution of Ca(OH)□, 10% by weight] for several weeks.
2. **Demineralization:** The thoroughly washed and defatted crushed bones are subjected to demineralization by soaking them in a hydrochloric acid (HCl) solution with a concentration ranging from 4 to 10% for a duration of 1 to 2 days. Following demineralization, the bones are washed with water to yield a clean, soft stock of ossein.

3. **Extraction or cooking:** Controlled hydrolysis is employed to extract various grades of gelatine. The extraction process is conducted in multiple runs, each lasting 3-5 hours, with progressively increasing temperatures.

   a) An extraction temperature of 60°C produces the highest quality gelatine.

   b) An extraction temperature of 65-70°C results in medium quality gelatine.

   c) An extraction temperature of 80°C yields low quality gelatine.

   d) An extraction temperature of 100°C (boiling) produces glue.

   The remaining residue is pressed and dried for utilization as livestock feed or fertilizer.

4. **Filtration:** To improve clarity, the liquid extracts of the higher grades undergo pressure filtration.

5. **Concentration:** The filtered liquid extracts are concentrated through vacuum evaporation, leading to gelatine with a concentration of 30-40%.

6. **Drying:** The concentrated gelatine is spread into a thin layer on a large, steam-heated drum. It is dried within a few minutes and subsequently removed with a knife. Depending on industry requirements, it can be sold as sheets, broken into flakes, or powdered.

Hide and skin trimmings, ear pieces, poultry skin, sinews, tendons, horn pith, and casings are utilized to produce glue and are collectively referred to as glue stock. Zinc sulphate is incorporated as a preservative to extend its shelf life.

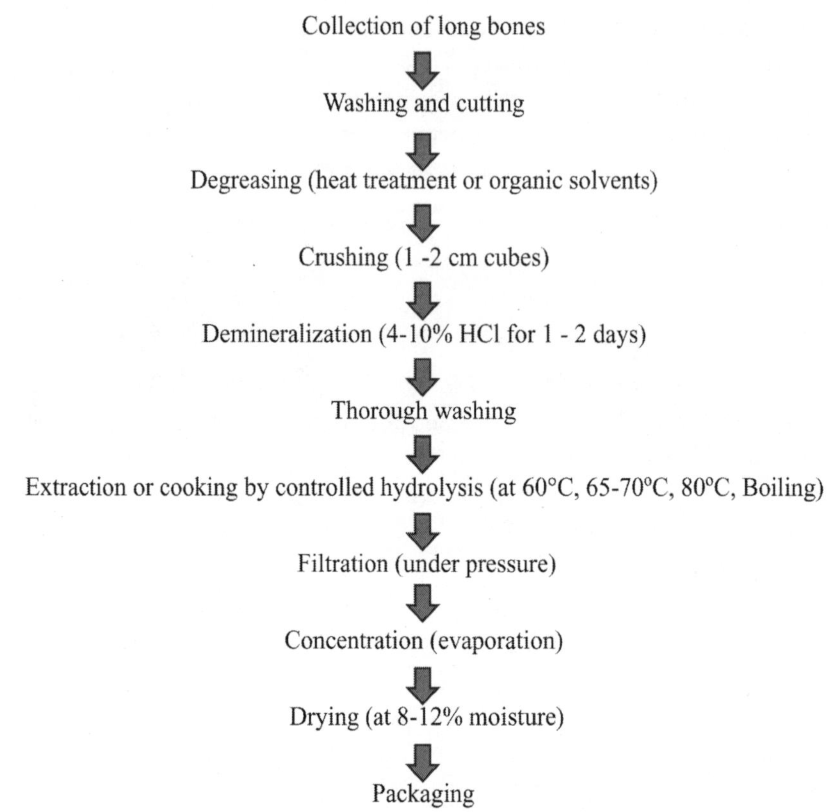

Figure 13: Flow diagram for the preparation of gelatine

B. **Bone meal:** Bone fragments measuring less than 2 mm are categorized as bone meal, which acts as a significant phosphate additive in animal feed. Insufficient phosphorus levels can lead to health issues in animals, such as osteophagia, osteoporosis, and rickets. Green bones, sourced directly from slaughterhouses, are rich in moisture, protein, and fat. Conversely, desert bones, which have been subjected to bacterial, atmospheric, and insect degradation over time, lose their meat, tendons, protein, fat, and moisture content. These bones are lightweight and are primarily composed of calcium, phosphorus, and dried ossein.

Collecting and processing desert bones into bone meal is not only economically feasible but also advantageous from a health standpoint. This practice can create job opportunities for impoverished and uneducated individuals while enhancing the health of livestock. It is crucial to sterilize these bones.

The preparation of bone meal mainly involves a wet rendering technique. In this process, bones are subjected to cooking in a vertical wet rendering cylindrical cooker at a steam pressure of 40 psi for a duration of 4 to 6 hours or at 60 psi for 2 hours. Following the cooking phase, the adhered water and fat are removed through gravity decantation, allowing the mixture to settle for approximately one hour. The extracted fat is then moved to a fat settling tank, and the adhered water is drained away. The remaining cooked bones are subsequently dried in a dry renderer or pan dryer to eliminate any residual moisture. Sun drying or utilizing solar heaters represents a more economical approach for this drying procedure.

Collection and cleaning of bones

⇓

Sterilization (Double jacketed bone digester, steam at 60 psi for 2 hours)

⇓

Draining

⇓

Cooling

⇓

Drying under sun

⇓

Milling (pulverizing)

⇓

Packaging

Figure 14: Flow diagram for the preparation of bone meal

The production of bone meal yields about one-third of the mass of the original bones, leading to a ratio of 1:3. The quality of bone meal is mainly assessed based on its phosphorus and calcium levels, which should optimally be in a 1:2 ratio. The standard composition of bone meal is generally presented in the table below:

Table 12: Composition of bone meal

Particulars	**Availability**
Calcium	30.5%
Phosphorus	15.5%
Protein	7%
Fat	1%

C. **Cattle lick:** Cattle lick serves as a supplementary feed formulated to supply vital minerals and nutrients that may be absent from the regular diets of cattle, thereby improving their health and productivity. It is generally composed of the following ingredients:

- 66 parts bone meal,
- 33 parts red oxide salt (which contains iron), and
- 1 part each of copper sulphate, potassium or sodium iodide, cobalt nitrite, sulphate or chloride, along with other trace elements.

This specific formulation guarantees a well-rounded intake of essential minerals and trace elements necessary for optimal cattle health.

D. **Bone ash or calcined bone:** Bone ash is mainly made up of calcium phosphate, which is obtained from the incineration or calcination of animal bones.

Preparation

1. **Collection and cleaning:** Animal bones, usually sourced from cattle, are gathered and meticulously cleaned to eliminate any remaining tissues and fats.
2. **Calcination:** The cleaned bones are subjected to high temperatures in a furnace. This procedure, known as calcination, causes the bones to decompose due to the heat.
3. **Grinding:** Following calcination, the bones are allowed to cool and are then ground into a fine powder, which is referred to as bone ash.

Uses

1. **Fertilizer:** Bone ash is abundant in phosphorus and calcium, rendering it a valuable fertilizer. It supplies essential nutrients that foster plant growth, particularly in crops that necessitate phosphorus for flowering and fruiting.
2. **Ceramics:** Bone ash is utilized in the production of ceramics, especially within the porcelain industry. It is blended with clay to create bone china, which is esteemed for its translucency, whiteness, and durability.
3. **Metallurgy:** In metallurgical contexts, bone ash functions as a fluxing agent. It aids in the removal of impurities from metal ores during smelting and promotes the formation of slag.
4. **Glass making:** Bone ash is incorporated in glass production to improve the clarity and strength of the final product. It can serve as a clarifying agent, enhancing the transparency of glass items.
5. **Animal feed:** In certain instances, bone ash may serve as a calcium supplement in animal feed, although this particular use is less prevalent compared to other applications.

E. **Chicken bone liquid concentrate:** Fresh chicken bones, typically remnants or trimmings from poultry processing facilities, are utilized as the primary material for producing chicken bone liquid concentrate.

Preparation

1. **Crushing or grinding:** Chicken bones are crushed or ground into smaller fragments to enhance the surface area.
2. **Boiling:** The crushed bones are boiled in water for a duration of 8 to 12 hours to extract flavours, nutrients, and minerals.
3. **Skimming:** During the boiling process, impurities, foam, and fat ascend to the surface and are periodically skimmed off to clarify the liquid.
4. **Cooling:** After cooking, the liquid is permitted to cool to room temperature.
5. **Fat skimming:** The solidified fat on the surface is removed to decrease fat content.
6. **Straining:** The liquid is strained through a fine mesh sieve or cheesecloth to eliminate any remaining solids, resulting in a clear concentrate.

The final concentrate generally comprises approximately 5% solids, which include dissolved minerals, proteins, and other nutrients. The concentrated liquid functions as a base or flavour enhancer in a variety of dishes, contributing depth and umami to soups, stews, sauces, and other culinary creations. It is also rich in beneficial nutrients such as calcium and collagen derived from the bones. The liquid concentrate can be stored in the refrigerator for several days or frozen for extended preservation, and it should be kept in airtight containers to ensure freshness and prevent contamination. Additional components like herbs, spices, or vegetables may be incorporated during cooking to further elevate the flavour. In Asian cuisine, such bone-based concentrates are highly esteemed for their rich and savoury tastes.

Utilization of Horns

The horns of animals exhibit a range of shapes, sizes, and colours that are influenced by species, breed, sex, and other variables. They are composed of two principal components: horn pith and horn proper. The horn pith serves as the core or inner section and is an important raw material for the production of gelatine or bone meal. Initially, horns are scalded in water at a temperature of 60°C to facilitate softening, after which they are mechanically divided into horn pith and horn proper. The outer section, known as horn proper, is predominantly composed of keratin. This material can be utilized to create a variety of products, including buttons, combs, handles, and ornamental items, through controlled processes of heating, pressing, and cutting.

Figure 14: Horns of different animal as interior decoration

Utilization of Hoofs (Shine Bones)

Hooves, similar to horns, serve comparable functions. They are detached from the feet through steaming and soaking in warm water. Following meticulous drying (ensuring avoidance of direct heat or sunlight), they are classified as white, striped, or black, with white hooves commanding the highest market value. Hoofs are employed in the production of neat's foot oil and hoof meal.

A. **Neat's foot oil:** The production of neat's foot oil involves the removal, drying, and pulverization of the hoof's outer covering for use as fertilizer or in crafting artifacts. Shin bones from hooves are utilized to create neat's foot oil, a pale-yellow liquid that remains fluid even at freezing temperatures. Despite its high market value, the yield per animal is quite limited. Neat's foot oil is predominantly manufactured near large slaughterhouses and contains a significant amount of oleic acid (60%), rendering it useful for lubricating delicate machinery, tanning leather, and formulating wound-healing ointments. Currently, there are no commercial entities in India that produce this oil. The processing of neat's foot oil consists of the following stages:

1. **Collection and cleaning:** The feet are severed from freshly slaughtered animals and thoroughly washed to eliminate blood and dirt.

2. **Scalding:** The feet and hooves are submerged in boiling water, and the hoof shells are extracted using a hammer, revealing the shin bones.
3. **Cooking or extraction:** Shin bones are boiled in water at 85°C for approximately 8 hours in an open tank or large kettle, allowing the oil to rise to the surface.
4. **Purification:** The floating neat's foot oil is moved to another kettle, reheated at 85°C for 8 hours, impurities are permitted to settle for 2 hours, and subsequently, the oil is filtered.
5. **Dehydration:** The oil undergoes dehydration by being heated to 100°C for around 2 hours.
6. **Packaging and storage:** The oil is placed in appropriate containers and stored in a dark environment at room temperature.

B. **Horn and hoof meal:** A rendering facility that is employed for the production of bone meal can also be intermittently utilized to create horn and hoof meal, which acts as a superior nitrogen-rich fertilizer. The horns and hooves from cattle, sheep, goats, and pigs are subjected to processing in the renderer for a duration of 8 hours, after which they are dried and ground into a fine powder. Despite the fact that this meal is not incorporated as a supplement in livestock feed due to its lack of palatability and low digestibility, it greatly enhances soil fertility owing to its elevated nitrogen levels.

6

Utilization of Intestine

The intestines of domesticated animals, which are a byproduct of slaughterhouses, are employed in numerous applications. Intestines that function as food containers are referred to casings. Non-consumable uses encompass catgut, strings for musical instruments and rackets, along with various other specialized applications. Additionally, intestines can be ingested following meticulous cleaning and cooking with spices. Among these applications, sausage casings yield the highest profit. Different nations exhibit distinct preferences for sausages, and these preferences frequently determine the varieties of casings utilized.

Example:

Cocktail sausages : Weighing <15 grams are filled in casings of small diameter.

Hotdogs : Filled in sheep and goat casing.

Mortadella : Weighing about 10 kg, filled in beef bladder.

Bologna : Filled in beef bung, middles or rounds.

Blood sausages : Generally, filled into processed pig stomach.

In addition to intestines, casings can also be sourced from weasands (oesophagus), urinary bladders, stomachs, and rectums. Although artificial casings made from plastic materials and hydro-cellulose are prevalent in Western nations, natural animal casings continue to be in high demand due to their distinctive properties. Natural casings, which are obtained from the intestines or other parts of animals such as sheep, pigs, or cows, undergo processing to become edible and are utilized in the production of sausages and other processed meats. They are esteemed for their elasticity, enabling them to conform to the shape of the sausage filling, and they play a crucial role in enhancing the flavour and texture of the final product. Natural casings are frequently favored for traditional or artisanal sausages because of their authentic origin and their capacity to improve the taste and visual appeal of the sausages.

Artificial casings are manufactured from synthetic materials like cellulose, collagen, or plastic (e.g., polyamide) to mimic the properties of natural casings. These materials can be designed to possess specific characteristics such as uniform size, strength, and ease of removal post-cooking. Artificial casings are extensively employed in large-scale industrial sausage production owing to their consistent quality and cost-effectiveness. They guarantee uniformity in product appearance and performance. Furthermore, artificial casings are often selected in vegetarian or halal/kosher food production contexts where the use of animal-derived products is limited.

Table 13: Comparison between natural and artificial casing

Property	Natural casing	Artificial casing	
		Collagen	**Cellulose**
Strength	Variable	High and uniform	Good and uniform
Length	Variable	Standardized	Standardized
Cost	Low	High	Low
Diameter	Variable	Uniform	Uniform
Holes	Suspected	None	None
Smoke penetration	Best	Very good	Good
Presentation	Salted	Dried	Dried
Printability	None	Limited	Best
Refrigeration storage	Yes	Yes	No
Shelf life	Limited	Fair	Very good
Microbial quality	Suspected	Good	Good
Tenderness	Most tender	Less tender	Peeled
Breakage during processing	Most likely	Less likely	Least likely
Machinability	Least	Less	Best

India exports salted sheep casings to various developed nations, such as Japan, Germany, Spain, Italy, Switzerland, and Denmark, after undergoing inspection by the Directorate of Marketing and Inspection in accordance with the Animal Casing and Marketing Rules of 1964. This is due to the fact that numerous countries find it challenging to satisfy their demand for small-calibre casings. Furthermore, dried cattle casings are mainly exported to Gulf nations.

Histology of Intestine

When examined under a microscope, the intestines (from the innermost to the outermost layer) are comprised of four distinct layers: the mucosa, submucosa, muscularis externa, and serosa.

Mucosa: This innermost layer, also referred to as the intestinal lumen, is made up of large aggregates of epithelial cells and smooth muscle cells, accompanied by a limited quantity of connective tissue, lymphatic cells, and dispersed blood vessels.

Submucosa: The second layer from the interior, it is predominantly formed of collagen bundles and some elastic fibers, imparting a greyish-yellow hue. This layer houses numerous blood vessels and adipose tissue interspersed within the fibrous matrix.

Muscular coat: This layer is composed of two sub-layers (longitudinal and circular) consisting of smooth muscle cells, elastin, and collagen fibers. These muscular layers enable the movement and contraction of the intestine.

Serosa: The outermost layer, which is thinner in comparison to the other layers. It is chiefly made up of collagen and elastic fibers, along with loose connective tissue cells.

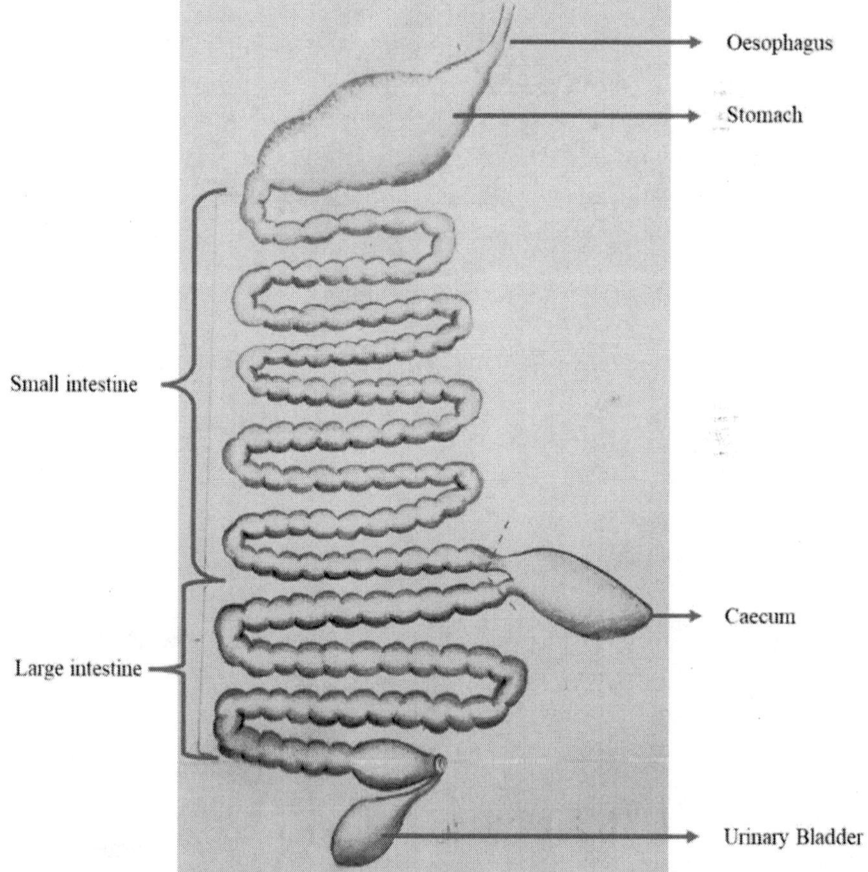

Figure 15: The alimentary tract and urinary bladder

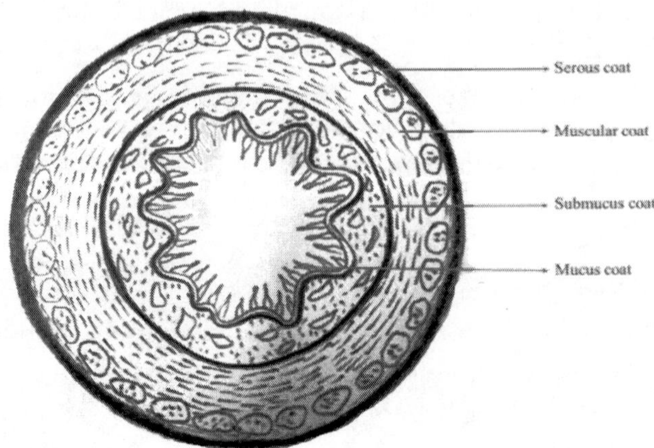

Figure 16: Histology of intestine

High-quality casings are meticulously crafted from the middle submucosa layer of the intestine. This layer, characterized by a greyish-yellow or red hue, contains a robust network of connective tissue interlaced with elastic fibers. Owing to its intricate structure, flexibility, elasticity, and durability, the submucosa is perfectly suited for the production of sausage casings. Its ability to contract and expand without fracturing renders it especially appropriates for this application.

Terminology

Commercial terms under which the casings after processing are sold differ from the anatomical names. Some of them are listed below:

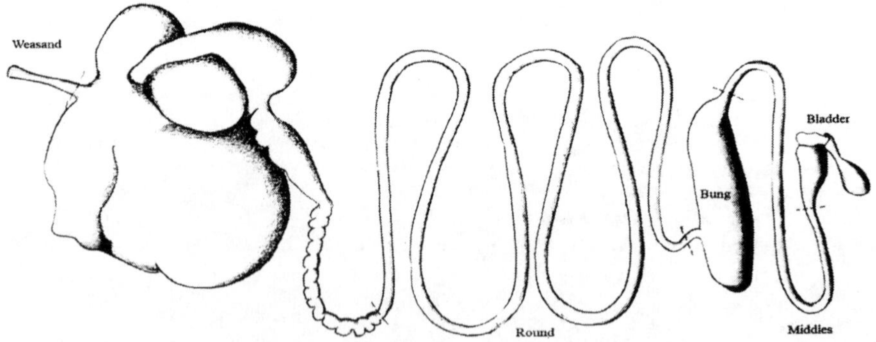

Figure 17: Different parts of cattle intestinal tract used as casings

Casings: Casings originate from the intestines or other parts of animals, typically from pigs, goats, sheep, or cattle. They undergo cleaning and processing before being utilized to encase sausage fillings.

Rounds: These are casings derived from the small intestines of sheep, goats, and pigs.

Runners: These casings are sourced from the small intestines of cattle.

Middles: These are casings obtained from the large intestines of both cattle and pigs.

Beef bung: This term refers to casings made from the blind gut or caecum of cattle.

Hog bung: Hog bung denotes the terminal segment of the intestinal tract of pig, encompassing approximately 5-6 feet of intestines that extend from the anus. It mainly comprises the pig's rectum and large intestine.

Weasand: These casings are prepared from the oesophagus.

Bladder: Casings derived from the urinary bladder.

Stomachs/maws: These casings are made from cleaned and sealed hog stomachs.

Small casings: These are casings sourced from the small intestines of hogs.

Chitterlings or black gut: These casings are prepared from a portion of the large intestines of hogs.

Sheep casings: These casings are derived from the small intestines of sheep.

Goat casings: These casings are sourced from the small intestines of goats.

Sheep and goat casings are typically stored in a wet and salted condition, while cattle casings are usually dried during processing and treated with insecticide for protection during storage. Casings are measured in hanks.

Salted sheep/goat casings: 1 hank equals 92 meters in length.

Dried cattle casing: 1 hank equals 180 meters for runners or 90 meters for middles.

Table 14: Approximate length of casings from animals

Species	Type	Length
Cattle	Runner	25-40 meters
	Middle	5.5-7.5 meters
	Bung	1-1.5 meters
	Weasand	5.5 meters
	Bladder	20-35 cm wide
Sheep	Round	27 meters
Goat	Round	22 meters

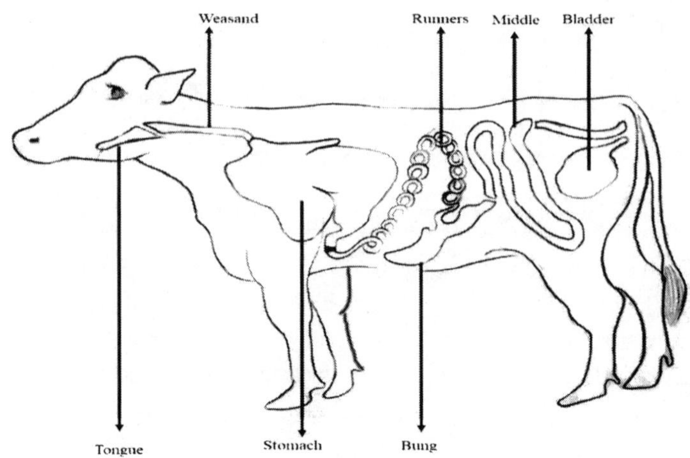

Figure 18: Important parts for casing preparation

Casing Processing Unit

In well-organized abattoirs that handle large volumes of slaughter, a specialized section for processing casings is frequently established to guarantee hygienic practices. This area is generally situated beneath the slaughter or dressing floor, where intestines are transported via chutes. The gut processing room must be adequately ventilated and provided with sufficient hot and cold potable water. It is furnished with machinery such as rollers and brushes, which serve to crush the mucous membrane, muscles, and fat, while simultaneously washing away scum and slime through a continuous water spray. It is essential to maintain a regulated temperature in the processing room to avert decomposition and fermentation. Alternatively, if required, guts may be cleaned and processed on a sanitized floor.

Processing of Casings

The majority of casings are derived from small intestines. For commercial purposes, the intestines of cattle, sheep, goats, and pigs are utilized in the production of casings. To ensure the production of high-quality natural casings, several critical factors must be taken into account:

1. **Healthy animals:** Intestines should be obtained from healthy animals that have undergone proper slaughtering and have successfully passed both ante mortem and post mortem inspections to confirm they are free from disease.
2. **Sound raw materials:** The intestines must be devoid of ulcers and any damage, particularly those inflicted by parasites, such as nodule-forming parasites, which can compromise the quality of the casings.

3. **Sanitary handling:** Intestines should be extracted and managed with rigorous sanitary measures. It is crucial to ensurc that thc casings are clean, exhibit a desirable colour, are of appropriate length, are free from unpleasant odours, and are adequately cured.
4. **Proper grading and packing:** Casings should be meticulously graded and packed to preserve their quality and appropriateness for use.

The following equipment is indispensable for the processing of casings

1. **Scrapers:** These are crafted from hard wood or plastic and are designed in a knife-like shape. Spoon scrapers, often made from bone or plastic, are commonly employed for the cleaning of casings.
2. **Gut cleaning unit:** This unit comprises three vessels; one for water, one for chilled water, and one for warm water. The third vessel is fitted with a sliming platform. These vessels are organized on a sliming table to streamline the cleaning process.
3. **Sliming table:** This table is made of steel and features a central drain that slopes slightly towards it. It is utilized for placing the tubs and for desliming the intestines. A separate tub is designated for the collection of slime.
4. **Table for grading, salting, and curing:** This table is employed for the subsequent processing stages, including grading, salting, and curing the casings.
5. **Skeining basket:** This tool is utilized to uniformly bundle or coil the casings into hanks for efficient storage and handling.

In addition to the equipment, an abundant supply of cold, clean water is crucial for the processing of casings. The equipment setup may vary between small processing units and large processing plants attached to major abattoirs.

A. Small processing unit

1. Scrapers made up of wooden or plastic knives
2. Gut cleaning table with a central drain pipe
3. Two stainless vessels, one for fermentation and the other with chilled water to collect cleaned intestines
4. Table for grading and salting
5. Storage containers

B. Large processing plant

- **For salted sheep casings**

1. Casing stripper with rubber rollers
2. Casing crusher with roller of bronze
3. Casing finisher with smooth and fine rollers
4. Casing flushing table with tap
5. Inspection and grading table

- **For dried cattle casings**
  1. Cattle casing stripper and fatter
  2. Casing turning tank
  3. Cattle casing crusher and slimmer
  4. Cattle casing cleaner
  5. Equipment for inflation and drying

Processing Protocol of Animal Casings

Casings should be processed as promptly as possible following slaughter and evisceration, utilizing either a laboratory processing method or an industrial processing method. The objective of both methods is to guarantee that the casings are clean, well-preserved, and appropriate for their designated use.

A. **Laboratory processing method:** A small quantity of animal intestines may be processed by adhering to the following steps in a laboratory setting. This procedure generally involves smaller-scale, manual techniques, typically conducted in a controlled environment. It emphasizes the maintenance of high quality through careful handling and cleaning protocols.

1. **Collection of intestines:** Post-slaughter, intestines are gathered from the slaughterhouse, emptied of their contents, and rinsed in clean water. They must be processed swiftly to avert bacterial decomposition and damage. Fat and other tissues are eliminated. The intestines are subsequently immersed in a solution containing 0.2% sodium pyrophosphate and 1% sodium chloride for one minute.
2. **Sliming:** The outer layers (serous, muscular, and mucosal) are stripped away to retain only the submucosa layer. This is accomplished by gently drawing the intestines over a small glass rod on a table, thereby pressing out the layers. This procedure is repeated until the casings appear clean and white.
3. **Washing and grading:** The casings are rinsed with water and then categorized based on size, colour, and any imperfections such as leaks or nodules.

4. **Curing and draining:** Casings are preserved by incorporating salt in wooden boxes with perforations to facilitate brine drainage. This process lasts approximately three days, after which excess salt is removed by shaking.
5. **Wet and dry salting:** In the wet salting method, the casings are submerged in a salt solution within barrels. In contrast, dry salting involves rubbing the casings with fine salt, amounting to about 40% of the casing's weight.
6. **Storage and transport:** Casings are stored in plastic coverings, bottles containing salt, or in tins or wooden boxes for transportation.

B. **Industrial processing method:** A significant volume of animal intestines can be processed through the following steps in a large processing facility. This approach is employed in extensive operations and utilizes automated machinery and procedures to enhance efficiency and uniformity. It generally encompasses mechanized cleaning, grading, and curing systems.

1. **Removal:** The intestine is meticulously extracted without contaminating adjacent areas or inflicting damage, secured with ligatures at both ends to contain its contents.
2. **Pulling/Running:** This stage entails the separation of mesentery and fat from the intestine. When performed manually, it is termed "pulling," while the use of a knife or tool is known as "running." Typically, a knife is employed for the thicker, tougher intestines of cattle.
3. **Chilling:** If immediate processing of the intestine is not feasible, it is stored at temperatures between 5°C and 6°C to avert bacterial deterioration.
4. **Stripping:** This step involves the extraction of the intestine's contents by applying pressure with hands, tools, or machines. It is most effective when conducted under running water to swiftly eliminate the contents. A straightforward technique is to pass the intestine between two fingers of one hand while applying gentle pressure, simultaneously pulling with the other hand.
5. **Flushing:** Water is forced through the intestine to expel any residual contents.
6. **Defatting:** This process entails the removal of fat from the intestine. For smaller animals, the fat is usually scraped off, whereas for beef intestines, it is extracted by pulling, tearing, or utilizing knives.

7. **Fermenting:** The intestines are submerged in warm water (70°C) for a duration of 1-2 hours to facilitate the loosening of layers through enzymatic and bacterial activity, enabling the removal of undesirable components. During winter, fermentation may occur at 20°C for 1-2 days. In summer, a small quantity of salt is added to decelerate the fermentation process. This technique is frequently applied to casings from sheep, goats, and pigs, but is generally unnecessary for cattle or buffalo intestines, which can be effectively cleaned using a knife.
8. **Turning:** Turning denotes the procedure of reversing the casing so that the inner surface is exposed outward. This step is crucial for effectively cleaning the intestines of cattle and buffalo. Typically, this process occurs in a warm water tank, where gravity-fed water from a tap assists in inverting the casing.
9. **Sliming:** Sliming involves the removal of the mucous lining, commonly known as slime, from the intestines. This is accomplished using a sliming stick or a plastic or wooden knife on a sliming board. The knife is positioned at a 30° angle and employed with gentle scraping motions to eliminate the mucous layer. The process is made easier if the casings are immersed in a sliming solution containing 0.2% sodium pyrophosphate and 1% sodium chloride for 10-15 minutes. Properly slimed casings should appear clean, white, and nearly transparent.
10. **Inspection:** The casings are examined for cleanliness, colour, odour, and any imperfections such as spots, holes, or parasites.
11. **Measuring:** The width of water-filled casings (for sheep and goats) or inflated casings (for cattle) is assessed using a calibrated casing gauge. The length is measured in hanks. Following measurement, the casings are classified based on their size and specifications.
12. **Preservation:** Sheep and goat casings are preserved in a wet state by salting with clean, medium-thick salt. They are manually turned and rubbed to ensure uniform salting, which improves their storage quality. Cattle casings are inflated with air, dried in a sheltered area, then deflated, pressed with a wooden roller, and cut to the required length.
13. **Packing:** Sheep and goat casings are packed in leak-proof metal cans, drums, barrels, or boxes containing brine. To prevent rust, the inner surfaces of metal drums are lined with plastic or cotton. Dried cattle casings are coiled on wooden wheels and stored in barrels with insecticide to ensure preservation.

Precautions

1. Casings should not be prepared in direct sunlight.
2. Casings should never be washed in rivers or streams.
3. Fresh and clean salt, free from stone and sand is essential for salting.

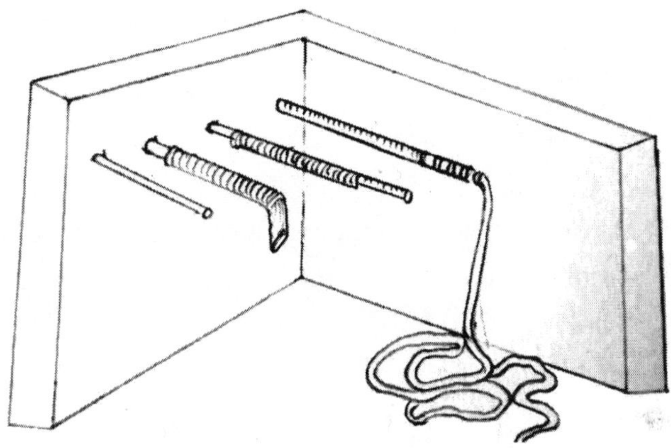

Figure 19: Turning beef gut inside out

Defects Generally Encountered in Indian Casings

1. **Dull hue:** Casings may exhibit a greyish or greenish tint rather than the intended white or milky appearance, signifying inadequate cleaning.
2. **Nodules:** The presence of nodules on the intestines may result from an infestation of esophagostomum (a type of roundworm) in cattle, sheep, goats, or pigs.
3. **Holes and lacerations:** These are frequently caused by rough handling or carelessness during the processing of the intestines.
4. **Salt burns:** This condition can occur when casings are stored in salt for an extended period, packed loosely with trapped air, or when the salt contains calcium and magnesium ions, leading to uneven burns.
5. **Defective grading:** Instances of defective grading may be noted in casings.
   - **a) Cicatrices:** Scars resulting from healed intestinal wounds.
   - **b) Domestics:** Small grease spots found in the casing.
   - **c) Kink:** A twisted loop present in the casing.
   - **d) Rust:** A black spot resulting from putrefaction due to bacterial or fungal activity.

Grading of Animal Casing

In accordance with the Bureau of Indian Standards IS: 1981-1962, which has also been adopted by Agmark as the Animal Casings Grading and Marking Rules of 1964, there are three classifications for sheep and goat casings, while four classifications are designated for dried cattle and buffalo casings.

A. **Salted sheep and goat casings:** The calibration ranges from 12 to 26 mm in increments of 2 mm, for instance, from 12 to 14 or from 14 to 16 mm is specified.

1. **Prime Quality (PQ) Grade or Grade I**

   a) The casings must

   i. Exhibit a natural colour throughout without any discoloration.

   ii. Be devoid of defects such as holes, blisters, lacerations, nodules, and cicatrices.

   iii. Remain intact and not be torn or lacerated.

   iv. Be free from salt burns, rust, domestic contaminants, black nodes, slime, mucus, dung, or mould infestations.

   v. Not rupture when filled with air or water to its normal capacity and slightly pressed.

   b) The rings or hanks must have been properly cured with common salt.

2. **Grade II:** This grade follows the specifications of Grade PQ, with the allowance of slight deviations in colour and/or strength and wall integrity. The material should be suitable for use in sausage preparation.

3. **Grade III:** This grade adheres to the standards of Grade II, with the exception that nodules are permissible.

   *For export purposes, an additional Grade X is available under Agmark for specific requirements agreed upon by the purchaser and exporter.

B. **Dried cattle and buffalo casings (Runners and Middles):** The calibrations in flat measurement range from 35 mm to 60 mm, with increments of either 2 or 5 mm specified.

1. **Prime Quality (PQ) grade: The casings must**

   i. Have a consistent natural colour, exhibit a lustrous appearance throughout, be devoid of spots or marks, and display no signs of discoloration.

   ii. Remain intact, without any tears or lacerations.

   iii. Be perfectly rolled.

   iv. Be free from black nodes, lacerations, or cicatrices.

2. **Grade I:** This grade adheres to the PQ standards, with the allowance for slight variations in colour and folds, as well as a few black nodes.
3. **Grade II:** Casings that do not meet the PQ or Grade I criteria due to defects in rolling, the presence of large black nodes, a rough texture, or a few streaks of fat.
4. **Grade III:** Short segments of any or all of the aforementioned grades.

7

Utilization of Glandular Byproducts

Animal organs and glands play a crucial role in both nutrition and medicine. These organs encompass the brain, heart, liver, lungs, spleen, tongue, pancreas, udder, stomach, uterus, testes, thymus, kidneys, parathyroid, adrenal glands, and ovaries. Some of these are consumed as food, while others find application in medical practices. Glands, constituting approximately 0.28% of an animal's total weight, are regarded as valuable by-products of meat production and are increasingly utilized in contemporary medicine for the extraction of beneficial substances. The secretions from both endocrine and exocrine glands are essential for the growth and functioning of the body, with hormones derived from these glands being employed to address deficiencies and serve as therapeutic agents.

1. **Human consumption:** The majority of glands are ingested alongside meat; however, in the case of cattle and buffalo, ovaries and testes are typically discarded, while other glands are consumed. In smaller animals, it is common to eat all glands with the exception of the ovaries. The liver is especially prized for its rich nutritional profile, being abundant in vitamins A, D, and B complex, which are advantageous for conditions such as poor vision, bone development, and anaemia.
2. **Medicinal uses:** Glands are responsible for producing hormones that offer significant medicinal advantages, including the treatment of anaemia, diabetes, and various other health issues. They also contribute to the development of secondary sexual characteristics in both males and females. Extracts derived from these glands are invaluable for restoring hormonal deficiencies within the body. The incorporation of animal glands into pharmaceutical manufacturing represents a notable advancement in modern science. In Western nations, there has been a concentrated effort on the efficient collection and utilization of these glands, acknowledging their economic significance for slaughterhouses and the meatpacking sector. Numerous life-saving medications that are currently imported could potentially be synthesized from glands, underscoring the necessity for improved domestic utilization of this

resource. Although synthetic alternatives are available, the untapped potential of readily accessible animal glands remains considerable.

Scope of Utilization of Glands in India

In India, the collection and application of animal glands is often overlooked, which is regrettable. The establishment of a dedicated industry for this purpose could yield significant advantages and ought to be regarded as a national priority. India relies heavily on imports of glandular products and unprocessed glands from nations such as the USA and the UK, primarily due to the absence of adequate facilities for gland collection and storage in local abattoirs. Currently, effective utilization is confined to major urban centres like Mumbai, Kolkata, Chennai, Delhi, Hyderabad, and Baroda, where animal slaughter rates are elevated, and glands such as the pituitary and pancreas could be utilized more efficiently if they were available in larger quantities. The conditions under which animals are slaughtered and processed in India impede proper gland collection. Abattoirs frequently lack cooling systems, and butchers often delay the removal of glands, resulting in the loss of valuable substances. To enhance the manufacturing of glandular products, it is essential to modernize abattoirs to facilitate proper extraction and storage. Furthermore, positioning pharmaceutical companies in proximity to abattoirs could optimize the process. Although the current slaughter rates in both municipal and rural abattoirs restrict gland collection, investing in appropriate equipment and personnel could render it feasible, provided there is a lucrative market. Nevertheless, the advent of synthetic alternatives for many glandular substances has reduced the potential for gland utilization.

Collection and Preservation of Glands

The following essential points must be adhered to for the effective collection and preservation of glands.

1. Glands should only be harvested from healthy animals.
2. During the excision of glands, it is crucial to avoid cutting the protective serous membrane to prevent the leakage of active principles and the risk of bacterial contamination. Incising the membrane also facilitates the entry of bacteria that are invariably present in the environment, leading to potential damage.
3. Glands must be extracted within 15-30 minutes post-slaughter to reduce exposure to elevated temperatures and moisture, particularly for smaller glands such as the suprarenal, parathyroid, and pituitary glands, as delays in removal can result in rapid degradation of hormones.

4. Glands should not be allowed to come into direct contact with water, as this can result in the leaching of active principles. After extraction, glands should not be washed; however, if necessary, they may be briefly rinsed with cold water to eliminate blood and visceral contents.
5. Glands should be immediately chilled in metal containers surrounded by crushed ice, ensuring they do not directly contact the ice. Smaller glands should be stored in glass or metal buckets equipped with drainage holes. These buckets should be filled with ice, ensuring that the glands do not touch the ice to prevent hormone extraction via water. Larger glands can be swiftly collected and placed directly into the cooler at regular intervals.
6. The subsequent step involves the cleaning and trimming of surrounding fat and connective tissue. Certain glands, such as the pituitary, ovary, and corpus luteum, require careful dissection into their individual components, necessitating a delicate and precise manual operation.
7. **Quick freezing:** The optimal method for preserving glands and inhibiting autolysis and bacterial proliferation is quick freezing or rapid freezing at temperatures between -18°C and -20°C. To facilitate quicker freezing, spread the glands out within an hour of removal. Ensure they are packed individually without any air pockets. After 48 hours, transfer the solidly frozen glands into covered containers to minimize air exposure and prevent freezer burn. Maintain them in a frozen state until they are processed by pharmaceutical manufacturers.
8. Glands such as the pancreas, pituitary, thyroid, and adrenal can be preserved chemically using solutions like acetone, phenol, or formalin. After 24 hours, it is advisable to transfer the glands to fresh acetone. The used acetone should be purified and reused. This method inactivates enzymes, prolongs shelf life, and eliminates fatty substances that may hinder protein extraction. It is essential to consult with manufacturers before employing chemical preservation methods to ensure that appropriate techniques are utilized.
9. **Preparation of acetone-dried powder from glands:** The acetone-dried powder can be prepared by adhering to the following steps.
   a) Collect the gland within 15-20 minutes post-slaughter.
   b) Eliminate connective tissues, blood vessels, and excess fat, then immerse the gland in four volumes of chilled acetone for three hours.
   c) Cut the glands into smaller pieces and soak them in three volumes of chilled acetone for an additional 2-3 hours.

d) Mince the treated tissue and process it further with three volumes of chilled acetone.

e) Dry the minced tissue and grind it into a fine powder.

10. **Vacuum drying:** The third method of preservation is vacuum drying. Glands that are dried under vacuum must be delivered to the manufacturer promptly to avert rancidity, which can diminish the efficacy of active compounds.

Despite the difficulties and low financial returns associated with the collection and preservation of glands, it is crucial to concentrate on enhancing these processes. Such improvements can positively impact public health and lessen the reliance on expensive imports of these substances.

Important Glands

1. **Pancreas:** The pancreas is a composite gland that is connected to the liver and surrounded by adipose tissue. It is chiefly recognized for its role in the production of insulin, an anti-diabetic hormone derived from the beta cells of the pancreas through the use of acidified methanol. In addition to insulin, which is widely utilized in pharmaceuticals, the pancreas synthesizes various enzymes that find applications in multiple industries, including tanneries. From one kilogram of fresh bovine or porcine pancreas, roughly 150 mg of crystalline insulin can be extracted, exhibiting an activity of 254 IU/mg. Other biochemical substances obtained from the pancreas comprise:

   a) Pancreatin- (Extract of pancreas)

   b) Trypsin

   c) Chymotrypsin

   d) Amylase

   e) Glucagon (β-cells)

   The pancreas harbours numerous enzymes that are beneficial in the tannery and cleaning sectors. To extend the preservation of these glands for up to a week, they may be treated with salt. The glands should be stored in drums with a layer of salt to ensure their integrity.

2. **Supra renal or adrenal:** The adrenal gland consists of two components: the outer cortex and the inner medulla. The cortex is responsible for the production of corticosteroids, which are utilized in the treatment of Addison's disease, the management of surgical shock, and other non-specific medical interventions. The medulla generates adrenaline (epinephrine) and noradrenaline. The yield of adrenaline is 0.2% based

on fresh weight and 1% based on dry weight, and it is extracted using either water or alcohol.

3. **Thyroid gland:** These two maroon-coloured glands are situated on either side of the trachea. Acetone-dried powder from the thyroid gland is employed to extract the thyroxine hormone using barium hydroxide, resulting in a yield of 0.08% of the gland's weight.
4. **Parathyroid gland:** Positioned adjacent to the thyroid gland, the parathyroid glands are responsible for the production of parathormone. This hormone plays a crucial role in preventing tetany and enhancing the rate of calcium excretion.
5. **Pituitary gland:** This diminutive gland necessitates meticulous collection. From the anterior lobe, the following hormones can be synthesized:

   a) Growth Hormone (GH)

   b) Follicle Stimulating Hormone (FSH)

   c) Luteinizing Hormone (LH)

   d) Adrenocorticotropic Hormone (ACTH)

   From the intermediate lobe, Melanocyte Stimulating Hormone (MSH) can be extracted.

   From the posterior lobe, the following hormones can be extracted:

   a) Oxytocin: Extracted using 2% acetic acid, it initiates uterine contractions and acts as a milk-ejecting factor.

   b) Vasopressin: Also known as antidiuretic hormone (ADH), it is a crucial vasoconstrictor and helps regulate water balance in the body.
6. **Ovaries:** Collected only from mature animals that have corpus lutea, the ovaries are irregular and lobulated, with large follicles resembling small grapes. Ovaries with cysts are discarded. From the ovaries, sex hormones such as oestrogen and progesterone are extracted.
7. **Testes:** Collected from mature animals that pass both antemortem and postmortem inspections. Testosterone, a sex hormone, is extracted from the testes. Additionally, the enzyme hyaluronidase, derived from bull or ram testes, is used as a spreading factor in drugs to enhance their effectiveness.
8. **Stomach glands**

   a) Rennin (Rennet): This enzyme is extracted from the lining of the fourth stomach (abomasum) of milk-fed and un-weaned buffalo or cow calves. The abomasum is removed, packed, and frozen without washing.

b) **Pepsin:** This enzyme comes from the mucosal lining of the hog stomach, which has distinct pink-red wrinkles and folds. The stomach is cut, and the glandular linings are separated and quickly frozen. Pepsin can be preserved in 1% sulfuric acid or as acetone-dried powder at room temperature, though very low temperatures are needed during preparation. Peptone can be made from the remaining part of the hog stomach.

9. **Gall bladder:** Connected to the liver, it holds bile, a viscous fluid that is dark golden-green in colour and possesses a bitter flavour. Bile is slightly alkaline (pH 7.15) and has a specific gravity of 1.025. When diluted in warm water (1:5 v/v), it can function as a detergent in slaughterhouses. Dried bile is utilized for medicinal purposes as a replacement for secretion deficiencies. One kilogram of bile is extracted from the gall bladders of six buffaloes or fifty-five sheep/goats and is preserved in a frozen state.

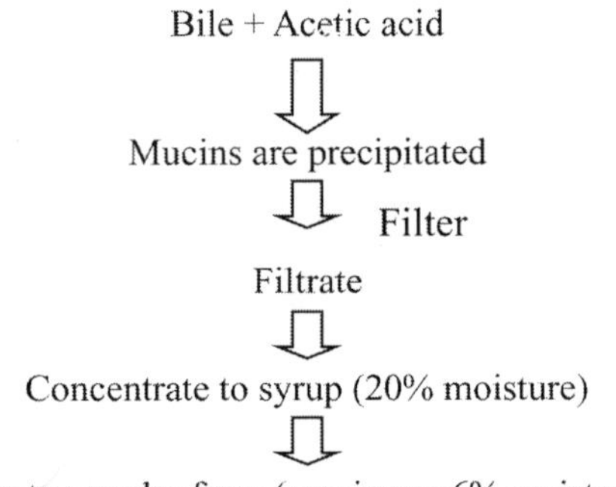

Figure 20: Manufacture of bile dried powder

Table 15: Important glands and organs of value

Gland	Active principal
Liver	Liver extract
Pancreas	Insulin, Pancreatin, Trypsin, Chymotrypsin, Carboxypeptidase, Lipase etc.
Pituitary - anterior lobe	ACTH, Prolactin, Growth hormone etc.
Pituitary - middle lobe	MSH
Pituitary - posterior lobe	Oxytocin and Vasopressin
Thyroid	Thyroxin
Adrenals	Adrenaline and Cortisones

Lungs	Heparin
Suckling calf stomach	Rennet and Pepsin
Gall bladder	Bile, Cholic acid, Cortisones, Bile salts, Cholesterol etc.
Brain	Cholesterol
Blood	Serum, Plasma, Fibrin, Haemoglobin etc.
Pineal gland	Melatonin
Testes	Testosterone
Ovaries	Oestradiol, Progesterone, Relaxin etc.

Some Extract From Glands

- Corpus luteum extract
- Liver extract
- Mammary gland extracts – source of substance, which stimulate the flow of milk and helps to control menstrual disorders.
- Brain extract
- Pancreatic extract
- Whole pituitary extract
- Anterior pituitary extract and posterior pituitary extract
- Testicular extract
- Thyroid extract
- Insulin
- Heparin (from lungs and liver of ox)

Preparation of Extract From Glands

1. **Preparation of insulin from pancreas**

Grind hard frozen glands in acid alcohol (acidified methanol) to extract insulin

⇩

Concentrate at temperature below 40°C

⇩

Separated by salting out, crude insulin goes to the top as a cake

⇩

Dissolve crude insulin in alcohol, then add diethyl ether, isoelectric precipitation pH 4.5

⇩

Crude insulin is purified by crystallization using zinc salt or picric acid

⇩

Insulin picrate can be stored for a long time

⇩

Insulin can be recovered by gel filtration

Figure 21: Flow diagram for preparation of insulin from pancreas

2. Preparation of heparin from cattle lungs

Minced lung tissue

⇩

Autolyse at 30°C for 24 hours; Toluene is added to avoid putrefaction

⇩

Add ammonium sulphate [$(NH_4)_2SO_4$] + NaOH mixture

⇩

Heat to 50°C, maintain for 1 hour for extraction

⇩

Precipitate with alcohol for 20 hours

⇩

Subject to enzymatic hydrolysis with trypsin centrifuge to gel clear solution

⇩

Precipitate crude heparin with 2 volumes of alcohol

⇩

Purify by gel filtration

⇩

Conduct sheep plasma assay to examine biological activity

Figure 22: Flow diagram for preparation of heparin from cattle lungs

3. Preparation of liver extract from liver (melt)

Raw minced liver

⇩

Suspend in water at 85°C

⇩

Digest at neutral pH with pepsin for 3 hours

⇩

Inactivate enzyme at 90°C for 30 minutes

⇩

Coagulate proteins

⇩

Digest in filter pressed

⇩

Clear filtrate is concentrated to syrupy consistency

⇩

Drying in vacuum to form powder

Figure 23: Flow diagram for preparation of liver extract

Future prospects: The modernization of slaughterhouses is leading to a greater utilization of glandular byproducts. Although there have been significant developments in synthetic alternatives, natural products continue to be more economically viable. Initiatives are in progress to create facilities for the prompt collection and immediate freezing of these glands, as well as to train staff for their preliminary processing. Additionally, research laboratories are focused on innovating new techniques and refining current technologies to boost the efficiency and effectiveness of glandular byproduct usage.

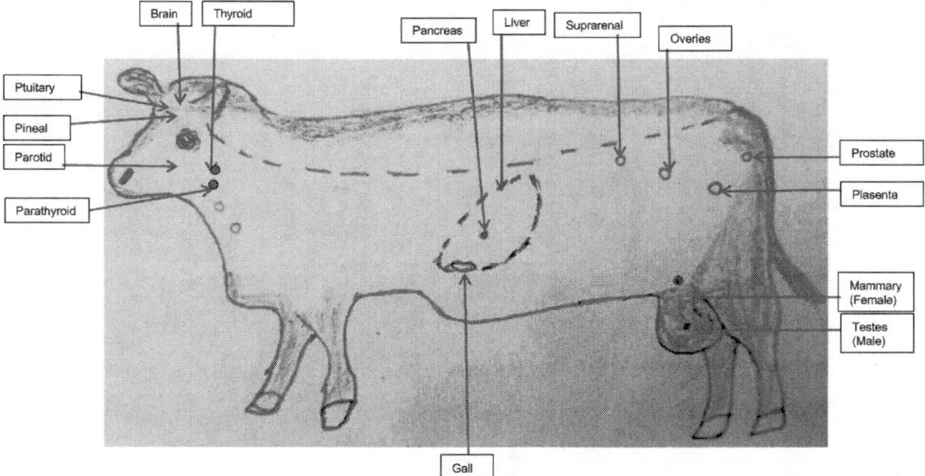

Figure 24: Positions of glands for pharmaceutical and other products

8

Harvesting and Processing of Hides and Skins

Historically, humans have utilized animal skins for various purposes including shelter, clothing, weapons, and food containers. In contemporary times, skins and hides significantly contribute to foreign exchange earnings within the animal husbandry sector and are widely employed by the leather industry. At slaughterhouses, hides and skins are generally collected and dispatched to tanneries on the same day. If a tannery is not in close proximity, hides are preserved through salting or curing prior to being transported in batches.

A skin (the outer covering) from a fully grown large animal is termed a hide; it is comparatively large, thick, and heavy, typically exceeding 30 lb (13.62 kg) for cattle or buffalo. Conversely, the outer covering of smaller animals such as sheep, goats, pigs, and young calves is known as skin. This type is smaller, thinner, and lighter. The term "slunk" refers to the skin of an unborn calf, which is often utilized for parchment, light suede, or drum skins. Calf skin is derived from young calves that have not yet reached maturity, and it is generally softer and more delicate than the hides of older animals. Its fine texture renders it highly sought after for premium leather goods. In contrast, kip skin is obtained from older but still immature calves. It is thicker and more robust than calf skin while being more refined than adult hides. Kip skin is recognized for its blend of durability and softness, making it ideal for long-lasting leather products.

Hides and skins constitute approximately 4-11% of an animal's live weight, influenced by factors such as species, age, breed, and health. In cattle, the average hide yield is about 7% of live weight, whereas in sheep and goats, the average skin yield is around 11% of live weight. Hides and skins rank among the most valuable by-products of animals and are transformed into leather through the tanning process. The trimmings are repurposed for the production of glue, artifacts, or as feed supplements.

Skins and hides are classified based on their origin; from slaughtered animals and fallen animals. For hides, 25% are sourced from slaughtered animals while 75% are derived from fallen animals. In contrast, for smaller animals, 80% of skins originate from slaughtered stock, whereas merely 20% are derived from fallen stock. Hides and skins obtained from slaughtered animals are typically of higher quality compared to those sourced from fallen animals. To maintain optimal quality, it is essential that hides and skins be flayed immediately following death or slaughter, as any delays may result in decomposition and a decline in quality. In India, hides are categorized as "Oxen" (which includes those from cows, bullocks, bulls, and calves) and "Buffs" (which refers to buffaloes of any kind).

Histological Structure of Hides and Skins

A cattle hide is mainly divided into three principal layers:

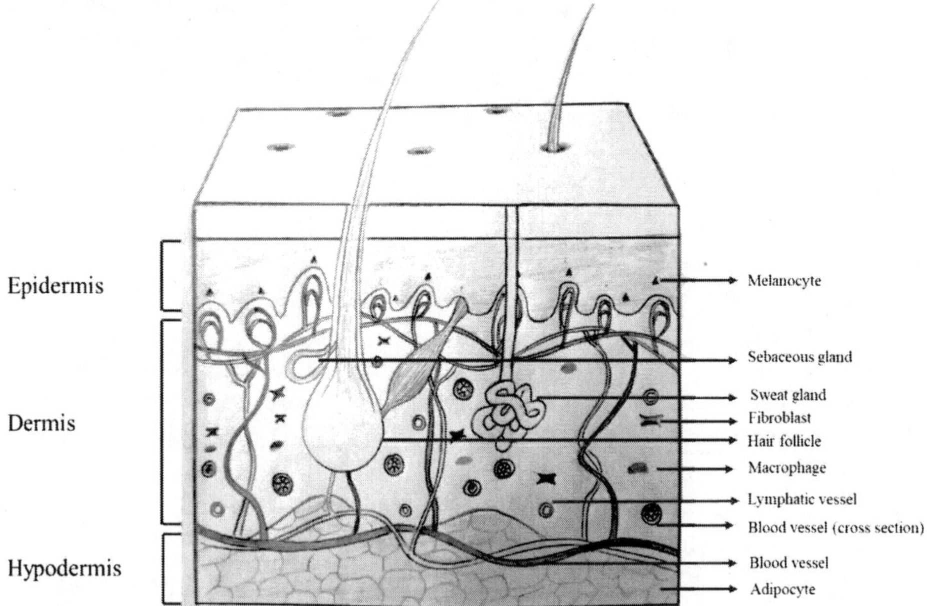

Figure 25: Histological structure of hide or skin

1. **Epidermis or outer layer:** The epidermis constitutes approximately 1-2% of the overall thickness of the skin and is primarily composed of epithelial cells. It does not possess its own blood vessels and instead derives its nourishment from the blood and lymph of the dermis. The epidermis is categorized into five distinct layers:

    a) Stratum corneum,

    b) Stratum lucidum,

    c) Stratum granulosum,

d) Stratum spinosum, and

e) Stratum basale.

In the process of leather production, the epidermis, along with hair, sebaceous glands, and sweat glands, is eliminated.

2. **Dermis or inner layer:** The dermis, often referred to as the corium, represents the principal layer of hides and skins, accounting for approximately 98% of the total skin thickness. It is predominantly made up of collagen fibers that are organized into bundles. The dermis is subdivided into two layers:

   a) Grain layer (Papillary layer): This layer comprises 10-25% of the total thickness and houses epidermal appendages such as hair roots, hair follicles, fat and sweat glands, as well as erector pili muscles. It plays a significant role in determining the appearance of the finished leather.

   b) Corium proper (Reticular layer): This layer constitutes 75-90% of the dermal thickness.

3. **Hypodermis or subcutis:** The subcutis is a layer of skin situated beneath the dermis. It is composed of loose connective tissue and contains adipose deposits. This layer serves to insulate the body and offers cushioning.

Classification of Hides and Skins

Hides and skins can be classified in various ways.

1. **According to weight:** The different types of hides and skins categorized by weight are illustrated in the table below.

Table 16: Different types of hides and skins according to weight

Origin	Weight of green hide in lb	Designation
Unborn calf	-	Slunk skin
Immature calf	9-15	Calf skin
Calf	15-25	Kip skin
Heifer	25-30	Heifer skin
Cow	>30	Cow hide
	30-53	Light cow hide
	>53	Heavy cow hide
Steer	32-48	Extremely light steer hide
	48-58	Light steer hide
	>58	Heavy steer hide
Bull	60-100	Bull hide

2. **In terms of branding:** Hides can be categorized into two types based on branding.

   Texas or Colorado hide: Hides that are branded on the buttocks or side are referred to as Texas or Colorado hides.

   Native hide: Hides that are unbranded are known as native hides.

3. **Based on the timing of flaying:** Hides can also be classified into two types according to the time of flaying.
   a) **Slaughtered hides:** Hides that are obtained through the flaying of slaughtered animals are termed slaughtered hides.
   b) **Fallen hide or Renderer hides or Murrain hides:** Hides that are acquired from fallen animals are referred to as fallen hides, renderer hides, or murrain hides.

4. **According to the quality of flaying:** Hides can be divided into two categories based on the quality of the flaying process.
   a) **Big packer hides:** Hides that are obtained through the flaying performed by highly skilled workers are known as big packer hides.
   b) **Small packer hides:** Hides that are obtained through the flaying conducted by less skilled workers are termed small packer hides.

5. **Classification of buffalo hides:** In India, buffalo hides are typically classified into the following categories:
   a) Buff (buffalo) calf skin
   b) Buff heifer hide
   c) Buff hide
   d) Buff bull hide

6. **Classification of goat and sheep skin:** In India, this classification is based on the weight and length of the hides.

Table 17: Classification of goat and sheep skin on the basis of weight and length

Class	Weight (kg)	Length (cm)	Region
Big goat skin	1.3	96.5 and above	Punjab, Kashmir-Punjab border, Orissa, Uttar Pradesh
Medium goat skin	0.91	96.4	Rajasthan, Maharashtra, Andhra Pradesh, Kerala, Karnataka
Small goat skin	0.86	76.2	Eastern part of India

*It may be noted that sheepskins are mostly of medium size without much regional variation.

Grading of Hides and Skins

FAO Expert, Aten (1995) suggested a useful system of grading hides/skins based on the degree of faults.

Table 18: Grading of hides/skins based on the degree of faults

Grades	Properties
First grade	• Shape or pattern regular and symmetrical • Minor scores and gouges may be over looked
Second grade	• Good shape or pattern • Reasonably free from knife damage i.e. up to 1/16th of area may show concentrated scores or gouges or 1/8th of area may show dispersed scores or gouges. • One or two cuts may be allowed on the edges of belly
Third grade	• Irregular shape or pattern • Up to 1/2 of the area damaged by knife showing cuts, scores or gouges.
Reject grade	• Irregular shape or pattern • Extensive damage of the back or butt.

Chemistry of Hides and Skins

The chemical composition of the skin or hide is as follows:

Table 19: The chemical composition of the skin or hide

Composition	Percentage
Water	65%
Protein	33%
Minerals	0.5%
Fat substance	2% (Cattle or calf), 2-10% (Goat), 5-3% (Sheep)

Animal skin is composed of different kinds of fats, such as triglycerides, phospholipids, cholesterol, and waxes. The mineral composition mainly includes phosphates, carbonates, sulphates, and chlorides of sodium, potassium, magnesium, and calcium. Moreover, around 80% of the dry matter in skin is made up of proteins.

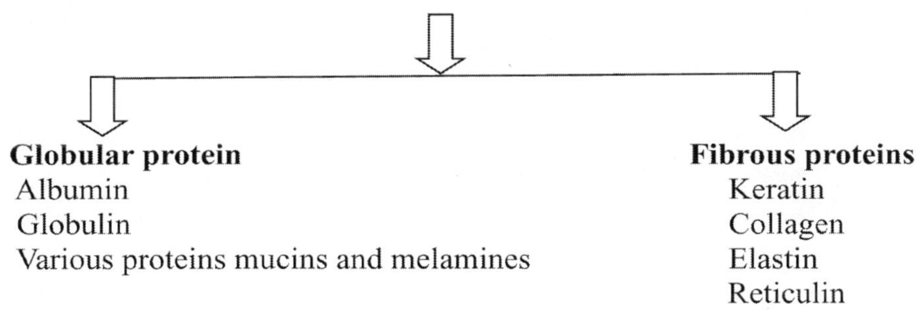

During the leather-making process, all proteins in the skin, except collagen, are removed. Collagen then combines with tanning agents to form the leather.

Flaying: Flaying, also referred to as skinning, is a meticulous procedure executed by adept workers to preserve the integrity of the skin. In modern large-scale animal slaughterhouses, buffalo and cattle are skinned on pritch plates, which are steel plates firmly secured to the ground. The animal is placed on the plate and restrained by a metal rod known as a pritch bar. This arrangement enables skilled workers to carry out the skinning process with efficiency. In the absence of pritch plates, a skinning cradle constructed from steel tubing or smooth timber rods can serve as an effective alternative.

India is recognized as the leading global producer and exporter of skins. These skins undergo processing into leather through the tanning process, while the trimmings are repurposed for the production of glue, artifacts, and feed supplements. Flaying is defined as the meticulous act of detaching the hide or skin from an animal following its slaughter or demise. Any imperfections such as cuts, tears, punctures, holes, or insect damage can significantly reduce the skin's value. Consequently, it is imperative that flaying is conducted by proficient butchers employing appropriate techniques. The quality of the skin is also affected by the animal's nutrition and management practices.

A. **Production of hides and skins:** It is crucial to exercise great care in the production of hides and skins derived from slaughtered and deceased animals. The flaying process should be executed promptly after the animal's death. Adhering to some of the following practices can contribute to the production of high-quality raw materials, namely skin and hide:

   a) **Feed and water management:** Animals should be withheld from feed and provided with sufficient potable water prior to slaughter. This increased water consumption facilitates the flaying of hides and skins by loosening the subcutaneous tissue.

   b) **Cold-water shower:** Before slaughter, animals should receive a cold-water shower. This practice diminishes blood flow to the peripheral tissues, thereby reducing the blood content in the subcutaneous layer. Such measures enhance the quality of hides and skins by improving their preservation and preventing complications such as 'veiny leather' and putrefaction.

   c) **Embrace contemporary flaying techniques:** Implement modern, scientifically-backed flaying practices with skilled personnel to guarantee superior outcomes. Well-executed opening lines (ripping lines) contribute to the production of uniformly shaped hides and skins. Eliminate non-leather elements such as ears, snouts, dew claws, and other appendages, as these can disrupt the tanning process and compromise the final quality of the leather.

d) **Enhance hygiene standards in slaughterhouses:** Numerous Indian slaughterhouses function under unsanitary conditions, resulting in rough treatment of animals and subsequent damage to the grain surface of hides and skins. Animals are frequently thrown onto filthy floors and dragged carelessly, inflicting harm on the hides. Following flaying, hides and skins may be left on contaminated surfaces or trampled with gastrointestinal contents to add weight. These unsanitary practices must be actively discouraged to ensure the production of higher quality hides and skins.

e) **Prompt flaying of deceased animals:** Skins from animals that have succumbed to natural causes, known as fallen hides or fallen skins, lose value if there is a considerable delay between death and flaying. To preserve the quality of these skins, they should be flayed as soon as possible after the animal's demise.

f) **Cleaning and preservation:** After flaying, hides and skins ought to be washed with cold water and a long-handled brush to eliminate dung and other impurities. Once cleaned, they are either dispatched directly to tanneries or preserved through curing.

B. **Scientific flaying of cattle and buffalo carcasses:** Flaying significantly influences the value of hides. To achieve optimal results, it is crucial to adhere to the following criteria.

1. Utilize specialized and extremely sharp knives with rounded blunt tips to avoid unwanted cuts.
2. Ensure proper grip of the knife.
3. Execute precise ripping lines/cuts.
4. Flay as soon as possible after the animal's death, as it becomes challenging to flay cold carcasses.
5. Maintain a stable position of the carcass during the flaying process.
6. Prevent contamination with manure and blood.
7. Whenever feasible, pull or beat the hide off.
8. Refrain from cutting or scoring the hide.
9. Utilize a flaying cradle.
10. Employ trained butchers.

C. **Flaying of buffaloes and cattle:** The following procedures are adhered to during the flaying operation of slaughtered cattle or buffaloes.

1. The hide is first opened from the neck or slaughter incision with a flaying knife, proceeding along the centre of the dewlap and belly to the midpoint of the tail.
2. Each leg is then encircled between the knee and the hoof joint, with the cut on the foreleg extending to the breastbone and the hind leg cut reaching the scrotum or udder, connecting with the initial longitudinal incision.
3. The hide is detached from the carcass by slicing from the lower breast to the neck on one side and to the navel on the opposite side, using a knife with caution.
4. The hind legs are suspended using a tail grip and gambrel linked to the hoisting apparatus, facilitating the removal of the hide from the tail.
5. The carcass is elevated further, and the hide is pulled off from the back to the hump, and subsequently to the shoulder and neck, with thick subcutaneous tissue being separated with a knife.
6. Ultimately, the hide is severed from the carcass behind the horns.
7. A properly flayed hide exhibits:
    a) Rounded rumps
    b) Uniform width from the centre line of the back to the belly edge on both sides
    c) Medium length in the shanks
    d) Consistent dewlap and
    e) Square outline

D. **Flaying of fallen cattle and buffaloes:** The process of flaying fallen cattle and buffaloes involves the use of hides from animals that have succumbed to natural causes such as aging or illness. These hides typically exhibit lower quality due to the presence of coagulated blood, which can lead to discoloration and potential decomposition, while the connective tissue may complicate the flaying process, resulting in cuts or gouges. Timely and meticulous flaying can mitigate damage.

The deceased animal is positioned on a wooden flaying cradle. This cradle can be easily assembled by placing two smooth wooden logs, each measuring 150 cm in length and 15 cm in diameter, spaced 30 cm apart. These logs are then interconnected with cross pieces of identical diameter and secured with nails from underneath.

E. **The flaying of sheep and goat skins:** The following technique is widely practiced in tropical and subtropical regions. It offers the benefit of producing skins with minimal cuts and superior preservation quality.

1. After the animal has been bled, create a small incision on the inner side of the hind leg, just above the hock joint.
2. Insert a narrow steel or smooth wooden rod into the incision and guide it beneath the skin towards the grain for approximately 45 cm to loosen the connective tissue, thereby reducing the reliance on a knife.
3. Withdraw the rod, make a hand cut around the incision, and introduce air using a bicycle or car foot pump.
4. The carcass will inflate like a balloon as a result of the air.
5. Elevate the carcass from the ground and suspend it by the hind leg on a hook or tripod.
6. Insert fingers and subsequently the fist into the hind leg incisions, pulling the skin downward, and utilize a knife only when absolutely necessary. Upon reaching the chest, grasp the skin with both hands and use your foot to push it off. This method may result in asymmetrical skins; however, careful flaying of a small section of the belly can assist in preventing this issue.
7. The following incisions are made in this technique:
   a) Circular cuts around the knee and hock joint.
   b) Cuts on the scrotum, udder, etc.
   c) A circular incision on the neck to sever the head skin.

Figure 26: General pattern of ripping lines

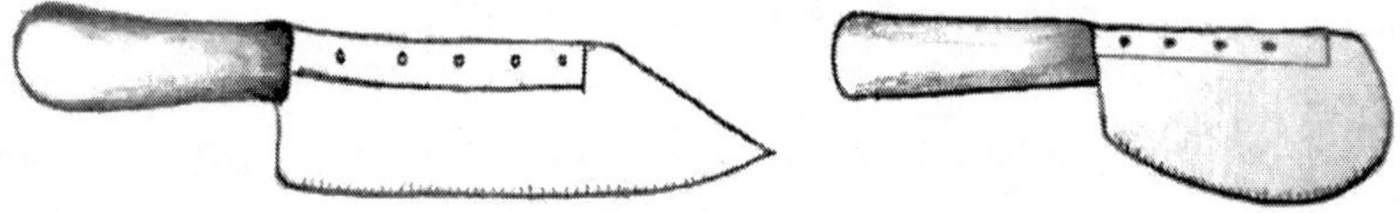

Figure 27: Ripping and flaying knives

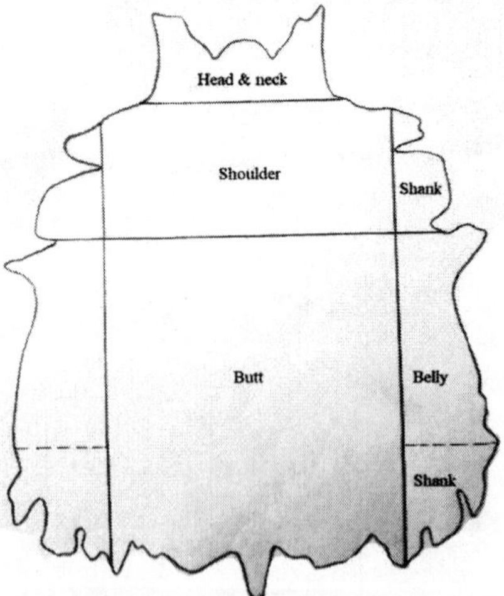

Figure 28: Principal areas of a hide

Defects of Hides and Skins

Defects can arise in hides and skins due to either diseases and insect infestations or improper handling practices.

1. **Defects resulting from diseases and insect infestations:** In tropical areas, animals are susceptible to a variety of parasitic, bacterial, and viral infections that can harm the hide and skin. The degree of damage is influenced by the duration and severity of the infection, which can range from minor imperfections to total destruction of the corium. Often, the damage is worsened by secondary infections and the animal's scratching or rubbing in response to itching. Malnutrition during droughts and insufficient green fodder, coupled with significant internal parasitic burdens, can exacerbate skin ailments. Furthermore, damage from ticks and the application of concentrated tick-killing medications can also lead to defects in hides and skins.

   a) **Follicular or demodectic mange:** A prevalent defect is attributed to the mite *Demodex folliculorum*, which causes follicular or demodectic mange. This parasite penetrates hair follicles, resulting in raised whitish lesions on the flesh side of the hide or skin. These lesions are often incorrectly labelled as pox marks in commercial contexts. True pox, which is viral in origin, produces very minor lesions primarily on the udder and inner thighs. If a vesicle ruptures, it may become infected due to significant irritation and subsequent scratching by the animal.

   b) **Sarcoptic and soroptic mange:** Sarcoptic and soroptic mange, commonly referred to as scab, are widespread due to warm climates and inadequate management practices. These parasites invade the spaces between the fibers of the skin's corium, resulting in rough, pitted leather with a compromised grain.

   c) **Dermatophilosis:** Dermatophilosis is a skin condition resulting from the bacterium *Dermatophilus congolensis*. This ailment predominantly impacts livestock, including cattle, sheep, and goats, particularly in tropical and subtropical areas and is known by various local names such as Krichi in Nigeria, Senkobo in Zimbabwe, and Uasin Gishu in Kenya. This disease can lead to minor skin inflammation affecting only the grain side of the leather or can cause extensive damage. In severe instances, scar tissue develops, and the hide thickens, resembling that of an elephant or rhinoceros, rendering it unsuitable for leather production.

d) **Nodular dermatitis:** This condition, likely induced by a viral agent, predominantly impacts goats and results in lesions that resemble those caused by Demodex mites. The specific viral agent responsible may differ, yet the Orf virus, which is also referred to as contagious ecthyma, is a frequent cause. Additionally, minor dermatological issues may stem from ringworm infections attributed to the Trichophyton genus. Although ringworm generally leads to the formation of round, hairless patches, excessive rubbing and subsequent infections can result in more profound lesions, thereby affecting the quality of the leather produced.

e) **Lumpy skin disease:** This disease, which is widespread in Africa, particularly south of the Sahara, inflicts considerable harm on cattle hides. It is characterized by the appearance of lumps, nodules, or circular patches that lose their natural texture, and it may also lead to button-like deformities that penetrate deeply into the corium.

f) **Photosensitization:** This phenomenon occurs when the consumption of specific plants or medications increases the skin's sensitivity to sunlight. It predominantly affects unpigmented skin areas, which are often observed in animals with white fur. Breeds such as Ayrshire and Friesian, originating from temperate climates, exhibit heightened vulnerability.

g) **Dermatitis:** This condition may arise from the application of highly concentrated acaricides or from their incorrect usage, impacting hides and skins in regions where tick management is practiced. In addition to the aforementioned diseases, various insects that infest living animals can also inflict damage on the skin.

h) **Warble fly (Heel fly or Grub):** Warble flies, which inflict considerable harm in temperate regions as opposed to tropical and subtropical areas, are classified under the family Hypodermidae. The three primary species that impact the northern boundaries of subtropical nations include *Hypoderma bovis*, *H. lineatum*, and *H. crossi*. These flies deposit their eggs on the fur of animals, and the larvae that emerge subsequently penetrate the skin. Following their migration through the host's body, they establish themselves in the subcutaneous tissue located on the back, resulting in the formation of noticeable lumps referred to as "grubs." The parasite breathes through small apertures, and after undergoing one or two moults, it transforms into a mature warble, descends to the ground, moults once more, and perpetuates the cycle. The degree of damage inflicted on hides and skins is contingent upon the stage of the cycle at the time of slaughter: open grubs create unhealed holes, while wounds lead to tissue damage.

i) **Tick damage:** The damage caused by ticks is prevalent, leaving hides and skins marked with pinhole spots where the ticks have attached themselves. Birds that feed on ticks can exacerbate the damage to the deeper layers, especially if the wounds become infected with bacteria.

j) **Other damages:** Lice, biting flies, and stinging insects may inflict minor blemishes on the surface of hides and skins. Nevertheless, insects such as the hide beetle (Dermestes) can cause substantial damage once the hides are detached from the animals. Both the beetle and its larvae are capable of rapidly consuming significant portions of a hide if it is not adequately safeguarded.

2. **Defects arising from improper handling**

a) **Fallen hides:** Hides obtained from animals that have succumbed to natural causes are termed fallen hides. These hides typically exhibit lower quality due to diminished hide substance resulting from fever or starvation. Furthermore, the process of flaying becomes more difficult as the subcutaneous connective tissues harden post-mortem.

b) **Ground drying:** This improper preservation technique leads to both visible and hidden damage, including hair slip, taint, and blisters, which may ultimately result in cracks forming in the hide.

c) **Smoke damage:** Damage to tanning materials may occur if they are stored in proximity to open flames. The heat generated by the fire can negatively impact the quality of the tanning agents.

d) **Used hides and skins:** Hides and skins frequently sustain damage when utilized as sleeping mats or clothing prior to sale. Such usage can lead to tearing, cracking, smoking, oil-tanning, or insect damage.

e) **Brand marks:** Negligent branding considerably reduces the value of hides. Hot irons should be applied only to less valuable areas, as branding for identification or disease control purposes diminishes the skin's worth.

f) **Wire damage:** Inserting wires into regions such as the brisket, dewlap, shoulder, or sides results in specific damage. This type of injury must be differentiated from scratches that occur when animals are confined in areas with barbed wire.

g) **Cracks:** Cracks form in over-dried hides when they are folded during transportation, leading to damage.

h) **Pressure sores:** This defect, also referred to as deductible gangrene, typically affects hides from animals that have experienced prolonged illness or starvation. It primarily impacts the hip and shoulder blade areas.

i) **Bad shape or patterns:** Damage inflicted by vermin such as hyenas, rats, and dogs, followed by the trimming of affected sections, results in an asymmetrical hide.

j) **Knife damage:** Damage to hides can occur from the use of inappropriate tools, such as sticking knives or those with spear-shaped blades, during the flaying process. Furthermore, carelessness, lack of skill, untrained speed, poor visibility, and flaying cold or inadequately prepared carcasses contribute to hide defects and damage.

k) **Bruises:** Bruises are frequently observed in hides either before arriving at the abattoir or during the slaughtering process.

l) **Insufficient bleeding:** In cases of incomplete bleeding, where blood vessels remain filled, the resulting hides are referred to as "veiny" leather, which is regarded as an undesirable defect.

m) **Dragged or rubbed grain:** Dragging carcasses over rough terrain or damaged concrete floors results in abrasion and damage to the hides.

n) **Thorn scratches:** During dry periods, animals that browse on bushes and thorn trees frequently incur scratches from thorns and damage to their hides due to grain exposure.

o) **Infrequent transport:** Defects and damages in transport often occur during the rainy season when hides are moved and stored for prolonged durations without sufficient protection from rain and insects. The rubbing and soiling that occur during transport can worsen the damage.

p) **Adulteration:** In numerous less-developed nations, hides are often coated or covered with mud, manure, or ashes, or left with an excessive amount of fleshing.

Alleviation of skin/hide imperfections: To mitigate these defects and damages, the following measures should be considered:

i. Programs aimed at enhancing the quality of hides should be developed.
ii. Training and demonstrations on proper flaying techniques should be provided.
iii. Abattoirs should be equipped with adequate facilities for the handling of hides and skins, as they generate significant revenue.
iv. Skinning cradles should be made available.

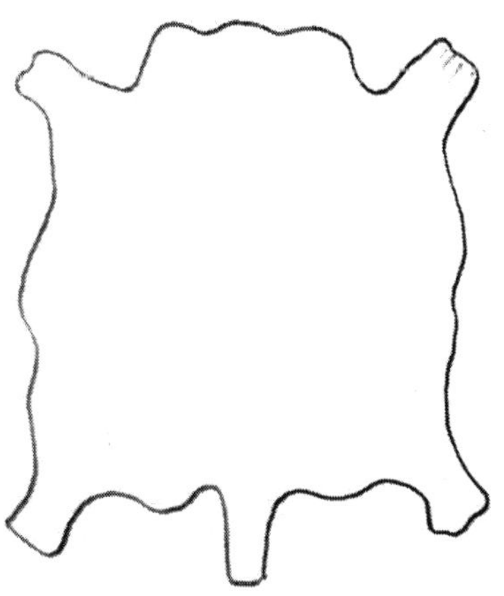

Figure 29: Properly flayed hide

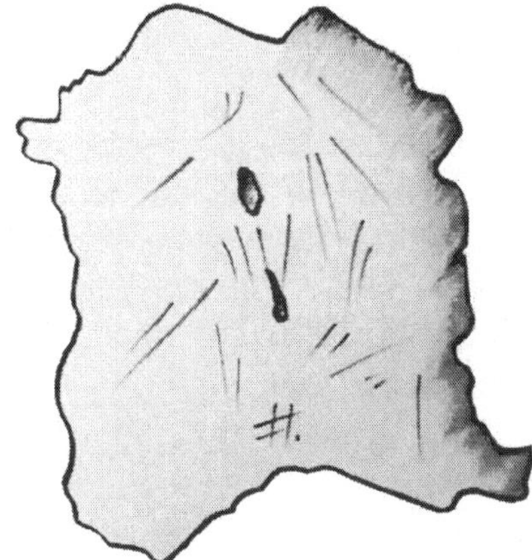

Figure 30: Improperly flayed hide

Care of Hides and Skins

To avoid defects, it is recommended to provide animals with cold-water showers or allow them to pass through cold water to eliminate dirt, dung, and other particles prior to stunning and slaughtering. This practice helps prevent the occurrence of 'veiny leather.' Following the flaying process, skins should be promptly opened and laid out on a concrete surface to cool. Subsequently, they must be thoroughly washed with water and a long-handled brush to eliminate dung, dirt, and blood. The removal of loose subcutaneous tissue, meat, and fat is achieved through a method known as fleshing. All sharp and irregular edges should be trimmed, the tail should be severed, and any portions below the knee and hock joints should be discarded. The skins are then either dispatched to a tannery or preserved using salt or sun-drying.

Preservation of Hides and Skins

Hides and skins are especially susceptible to spoilage due to their 60-65% water content, rendering them vulnerable to bacterial or enzymatic decomposition, particularly when tainted with blood and manure. To ensure their preservation, immediate actions are crucial following the flaying process. Start by washing to eliminate blood and dirt, utilizing scrubbing brushes on a platform beneath running cold water, and then position the hides on draining racks. Trim the hides to remove non-leather components and ensure they adhere to standard patterns. Ante-mortem care involves providing sufficient water and withholding feed for a minimum of 12 to 18 hours to facilitate flaying and minimize unnecessary cuts. Elevating the animal enhances bleeding and maintains skin cleanliness. For sheep and long-haired goats, it is advisable to wash only the flesh side of the skins.

"Green hides and skins" refer to those that have undergone the processes of flaying, fleshing, trimming, and washing, yet remain untanned. It is imperative that these are forwarded to tanners at the earliest opportunity. Should immediate delivery not be possible, it becomes essential to implement short-term preservation and appropriate storage techniques. Preservation generally entails the reduction of moisture through adequate air circulation, the application of salt, or ideally, a combination of both methods. The selection of a preservation technique is contingent upon environmental factors such as temperature, relative humidity, and the duration of storage.

Methods for preserving skins or hides: Elevated moisture levels and the presence of protein create a conducive environment for putrefactive bacteria, which can lead to the decay of skin and hide if not managed effectively. In the absence of measures to inhibit bacterial proliferation, hides may suffer deterioration, resulting in substandard leather quality or total spoilage. Early indicators of spoilage include patchy hair loss, unpleasant odours, and dark discolorations on the flesh side of the skin. The objective of preservation is to avert bacterial putrefaction and physical damage, such as wrinkles and insect bites, prior to the hides being dispatched to the tannery. This is accomplished through the processes of hide curing and air or sun drying.

1. **Hide curing:** In this method, drying is performed using uniform, finely ground salt, which should be applied immediately following the flaying process, utilizing one of the following three techniques:

   a) **Dry salting:** This technique aims to reduce the weight of hides for more convenient transportation while enhancing their preservation by preventing decay. The salt serves to inhibit saprophytic organisms, thereby aiding in the maintenance of the hides' condition. It is particularly beneficial for small-scale producers and those located far from major markets. In India, the practice of dry salting frequently employs khari salt (sodium sulphate), which is mixed with water in a ratio of 1 part salt to 2 parts water. The skins are washed, trimmed, and then rubbed with dry salt on the flesh side before being stacked in layers overnight. This method is prevalent in tropical regions. The initial salt packing assists in moisture removal, with any remaining moisture evaporating due to air exposure. Optimal storage conditions are approximately 15°C with 85-90% relative humidity, adequate ventilation, and a slatted platform. Hides should be stacked flesh side up with fine salt evenly distributed, corresponding to the weight of the hide. The stacking height should not exceed one meter to facilitate moisture drainage, ensuring that the hides maintain around 12% water content.

b) **Wet salting:** The process outlined is referred to as green salting. In this approach, freshly flayed hides and skins are placed on a sloped concave platform to encourage cooling. Salt is generously sprinkled on the flesh side and left overnight. For the curing process, a solution is created using 23 kg of pure salt and 62 kg of water for every 100 kg of hides. A saturated brine solution may also be utilized. Both rock salt and sea salt are appropriate for wet salting. The salt mixture can include:

- 100 parts salt + 2 parts naphthalene + 2 parts Na_2CO_3
- 100 parts salt + 1 part naphthalene + 1 part boric acid.

In the subsequent morning, the salt is taken away, and a new layer is applied, ensuring that thicker sections of the hide receive a greater amount of salt. Another hide is then positioned on top, with the grain side facing down, and the procedure is repeated. The hides are stacked in a pile until they are completely cured, with the piles being periodically tilted and rearranged to avoid heat accumulation, which could negatively impact quality. For the soaking process, hides are submerged in a pit that is 1.25 meters deep for several days, with the soaking duration varying from 2 days for fleshed hides to 2 weeks for un-fleshed hides. Following the soaking, the hides are placed on a slatted platform to drain, resulting in approximately 35% water content remaining.

c) **Brine curing:** Brine curing, also referred to as wet curing or immersion curing, entails soaking clean, trimmed hides in a saturated salt solution for a duration of up to 1-2 weeks. This process can be conducted in containers such as pots, vats, or pits that are around 4 feet deep. To maintain the saturation of the brine, additional salt is incorporated as necessary. After soaking overnight, the hides are turned to ensure uniform curing. Upon completion of the 1-2 week period, the hides are drained and packed in layers. This method enables the hides to absorb salt while losing moisture, thereby creating an environment that is unfavourable for the enzymes and bacteria present in the hides. Hides that have undergone brine curing can be stored without the need for additional salt and can remain in good condition for an extended period. Experience suggests that brine curing is more effective than traditional wet salting or other techniques. Furthermore, the preservative properties of salt can be enhanced by incorporating 2-3% sodium carbonate and 1% naphthalene based on the weight of the salt.

Advantages of brine curing

a) More rapid and uniform.
b) Provides greater protection when subjected to adverse conditions during storage and transportation.
c) Brine curing nearly eliminates salt stains during prolonged storage.
d) Does not necessitate any washing prior to soaking back in the tannery.
e) The soaking period is significantly shorter compared to that of wet salted hides.
f) Results in enhanced fullness and a more vibrant grain of leather.
g) Can conveniently substitute the traditional dry salt curing method.
h) Both the flesh and hair sides exhibit brightness and are remarkably free from dirt contamination, with no salt adhering.

2. **Air-drying/ sun drying:** Sun drying, which may involve ground drying or air drying, is frequently employed to preserve skins obtained from deceased animals in rural regions. To avoid damage and maintain strength, direct exposure to sunlight should be minimized. Instead, skins are stretched and secured to wooden or metal frames for drying in open air. The drying environment must be well-ventilated to promote adequate air circulation. This conventional technique is appropriate for areas with low relative humidity and is executed in three distinct methods:

   a) **Ground drying:** This approach entails stretching hides with the flesh side facing up on the ground. It is economical and suitable for rural settings with fallen hides; however, it may result in blemishes that become apparent post-tanning. Additionally, there is a risk of hair-slip and blistering due to premature decomposition of the epidermis and hair follicles. During summer, hides may wrinkle and only command half the price of salt-cured hides. Ground drying is generally less preferred as saprophytes can harm the grain surface, significantly diminishing the value of the skins and hides.

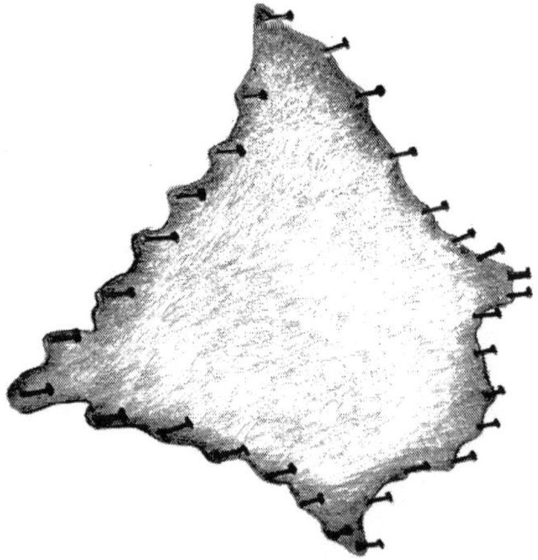

Figure 31: Ground drying

b) **Suspension drying:** Suspension drying represents an uncomplicated and economical technique, especially advantageous for tropical regions. Following the salting process, hides may be hung on extended bamboo poles to dry in sunlight, generally finishing within 2-3 days. In elevated temperatures (exceeding 115°F or 46°C), it is advisable to dry the hides in shaded areas to avoid deterioration in quality. Conversely, during cooler seasons, direct sunlight can be utilized, facilitating air movement and swift cooling. This approach not only lightens the hides but also diminishes transportation expenses.

Figure 32: Suspension drying

c) **Frame drying:** This technique entails the drying of skins on an inclined frame that is directed towards the sun. The frames may take the shape of hoops, tripods, or bamboo squares. Generally, after the salting process, skins are affixed to bamboo hoops or rope frames. Occasionally, skins are filled with straw and allowed to air-dry. The procedure consists of stretching the hides and skins lengthwise on bamboo frames and exposing them to gentle sunlight, with the flesh side oriented towards the sun. It is crucial to avoid rapid drying in extreme heat to prevent the development of a hard crust on both the flesh and grain surfaces. It is recommended to wash the hides prior to placing them on wooden frames secured with ropes. The hides should be secured lengthwise, ensuring that the sides are evenly suspended, while being cautious not to overstretch. The frames ought to be angled to prevent direct sunlight exposure, which can lead to the hides shrinking and crumpling, a condition referred to as 'crumpled hide,' which diminishes their value. Line drying is appropriate for sheep and goat skins, where they are laid out on horizontal cords with the flesh side facing upwards. In tent drying, hides are elevated above the ground in a tent-like configuration using cords or wires. Regardless of the drying method employed, the process typically requires a minimum of seven days.

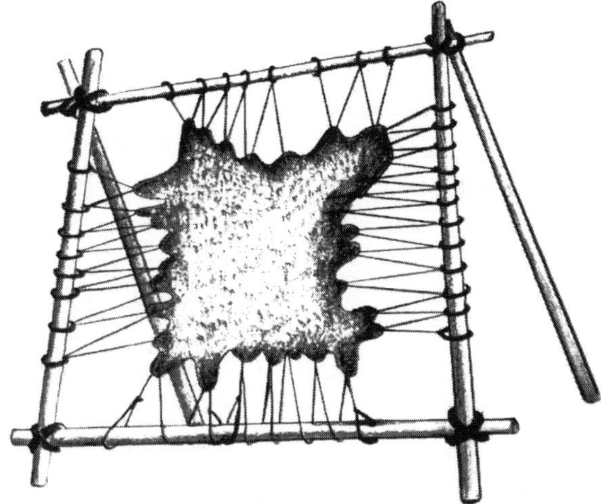

Figure 33: Frame drying

Advantages of air drying

a) The availability of air is abundant.
b) The removal of hair slip and decay.
c) Enhanced grading opportunities as any cuts, bruises, or parasitic damage can be easily detected.
d) Improved preservation quality.

e) Reduced transportation costs due to their lighter weight.

f) Corrosion is prevented.

Conditioning of Hides and Skins

The preserved hides and skins undergo initial conditioning at the tanneries prior to the tanning process, following these steps:

1. **Washing and soaking:** Hides are immersed for several hours in a solution of water mixed with zinc chloride, soda ash, and borax. This procedure eliminates salt, blood proteins, and lymph, enabling the hides to absorb water and restore their original shape and dimensions.
2. **Fleshing:** Excess flesh is removed on a convex wooden beam using a serrated knife to clean the hides effectively.
3. **Liming and dehairing:** A saturated lime solution, along with 0.1% sodium sulphide, is utilized to loosen and extract hair and epidermal cells.
4. **Washing and deliming:** Hides are rinsed with a mild acid to neutralize the lime. If necessary, gentle heat may be applied to facilitate this process.
5. **Bating:** Proteolytic enzymes (such as pancreatic juice combined with sawdust) are introduced at a pH of 8.5 to soften the pelt and enhance its pliability, preparing it for vegetable tanning.
6. **Pickling:** For chrome tanning, pelts are pickled in a mixture of 1% sulfuric acid and 10% salt in water, maintaining a pH of 2 to 2.5, for a duration of 2 to 3 hours.

Processing of Hides and Skins into Leather

The transformation of animal hide or skin into leather through chemical treatment and processing is referred to as tanning. This procedure increases the hide's resistance to decomposition and bacterial decay, especially when exposed to moisture, while also enhancing its physical characteristics such as tensile strength, flexibility, resilience, and abrasion resistance. The steps involved in the processing of hides and skins are as follows:

1. **Fleshing:** Fleshing entails scraping the flesh attached to the inner side of the skin with a serrated knife, which is essential for producing high-quality leather. Conducting fleshing prior to curing, particularly when the hide is warmer and more pliable, ensures a cleaner and more efficient process. A well-fleshed hide allows for quicker salt penetration and faster moisture removal, resulting in improved preservation and leather quality.

2. **Trimming:** The objective of trimming a hide is to eliminate non-essential portions and to standardize its form. This process entails the removal of ears, snouts, lips, scrotal sacs, udders, tails, head skin, and any irregular edges.
3. **Sorting:** Following the curing process, hides and skins are categorized according to sex, weight, and whether they bear branding. Branding reduces the value of the hide, as the branded section is not suitable for producing finished leather goods. The sorting procedure also includes grading the hides based on defects such as holes, cuts, visible grain problems, and evaluating the effectiveness of the curing process.
4. **Staining and storage of hides and skins:** The graded hides and skins receive treatment with approximately 1 pound of salt each to avert spoilage. Each hide is folded with the flesh side facing out and bundled together, secured with coloured ropes or tags for identification purposes. These bundles are then stored in warehouses. To guard against insect infestations, common insecticides employed include a 0.2% white arsenic solution mixed with soda, powdered 1,4-dichlorobenzene, or 0.5% Lindane.
5. **Soaking:** The soaking process reintroduces moisture lost during curing, which aids in managing bacterial proliferation. This procedure is conducted in half-round cylindrical vats where hides and skins are submerged in water, detergents, or disinfectants. The mixture is agitated with a paddle, and soaking generally lasts between 8 to 20 hours.
6. **Liming:** Hides are immersed in vats containing a saturated lime solution, which facilitates the separation of keratinous materials such as hair and epidermal cells, as well as loosening hair follicles. The dehairing process is predominantly chemical, but it also employs mechanical equipment. Common agents used in liming include hydrated lime [$Ca(OH)_2$], which helps to loosen hair follicles, and sodium sulphide (Na_2S), which dissolves the hair.
7. **Washing and deliming:** This phase eliminates alkaline dehairing agents. Limed hides are initially washed with water, followed by treatment with salts such as ammonium sulphate [$(NH_4)_2SO_2$] or ammonium chloride (NH_4Cl). These salts convert lime into soluble compounds, facilitating its removal by washing.
8. **Bating:** Bating refers to the treatment of delimed hides using proteolytic enzymes, such as trypsin derived from animal pancreas or wood. This procedure softens the hides and enhances their pliability. The treatment

occurs at a temperature range of 27°C to 30°C and a pH level of 8-9, lasting between 8 to 16 hours. Following the bating process, the hides are washed to eliminate any digested non-leather materials.

9. **Pickling:** The pickling process involves immersing the skins in a solution composed of sulfuric acid and salt. This method conditions the skins for tanning by establishing a consistent level of acidity, which is crucial for the effective absorption of tanning agents.

10. **Tanning:** Tanning is the procedure that transforms hides and skins into durable, non-putrescible leather while preserving their inherent structure. The leather produced through tanning acquires several advantageous properties, such as flexibility, resistance to heat and chemicals, abrasion resistance, and dimensional stability. It also retains its integrity through repeated cycles of wetting and drying. The tanning process entails treating conditioned hides and skins with a variety of agents, which may include organic, inorganic, or synthetic substances. Common tanning agents consist of vegetable tannins and basic chromium sulphate. Tanning can be broadly categorized into two primary types:

    a) **Vegetable tanning:** Vegetable tanning is a traditional technique that employs plant extracts from barks and other botanical parts to tan hides and skins. In India, frequently used plants include Avaram (*Cassia auriculata*), Babul (*Acacia arabia*), Myrabalan (*Terminalia chebula*), and Konnan (*Cassia fistula*). The process involves soaking de-limed pelts in increasingly concentrated tannin solutions within pits, transferring them from one pit to another to ensure uniform tanning. The use of resolving drums and concentrated tanning solutions accelerates the process. Vegetable-tanned leather is recognized for its durability, strength, and eco-friendliness, although the method is relatively time-intensive. This technique is particularly suitable for producing robust leather for items such as soles, belts, harnesses, and saddlery.

    b) **Chrome tanning:** Chrome tanning represents a contemporary technique recognized for yielding soft, supple, and resilient leather that permits air permeability and is more rapid than conventional methods. This process can be executed through two distinct approaches:

    - **Single bath process:** Basic chromium salt (such as chromic sulphate) is directly introduced into a solution, with its concentration gradually increased.

    - **Double bath process:** Initially, chromium salts are transformed into basic chromic sulphate via reactions with sugar and sulfuric acid. This resultant solution is subsequently applied to the hides within a drum. It is initially introduced at a pH of 2.8, which is later elevated to 3.5 to improve collagen bonding. This method entails the formation of cross-linkages between chrome ions and collagen, typically requiring 4-6 hours. Chrome-tanned leather is characterized by its flexibility, water resistance, and superior colour retention, rendering it highly sought after for a diverse array of leather products.

Post Tanning Operations

Post tanning operations are essential for achieving finished leather. Prior to the furnishing of leather, tanned leather undergoes a series of processes including setting, splitting and shaving, neutralizing and dyeing, fat liquoring and stuffing, setting out, pasting, conditioning, buffing, finishing, and grading. The sole leather is treated by applying oil to the grain surface followed by rolling under a heavy roller.

1. **Setting:** In the context of leather processing, setting can refer to various preliminary stages where the hide or skin is prepared for subsequent processing. This may include soaking the hides in water to soften them and eliminate impurities before any mechanical or chemical treatments commence.
2. **Splitting:** This procedure entails dividing the thick hide into layers. The uppermost layer, referred to as the grain layer, is utilized for higher-quality leather products. The lower layers, known as splits, are frequently employed for suede or lower-grade leather items.
3. **Shaving:** Following the splitting process, the leather may undergo further shaving to attain a consistent thickness and smooth surface.
4. **Neutralizing:** Following chemical treatments like tanning, the leather undergoes a neutralization process to stabilize its pH and cease the tanning procedure.
5. **Dyeing:** Leather can be coloured to attain the preferred hue. Various methods, including spraying, brushing, or immersion, can be utilized for dye application, depending on the leather type and the intended finish.
6. **Fat liquoring:** This procedure entails the application of oils, greases, or lubricants to the leather to enhance its flexibility, softness, and resistance to water. Additionally, it boosts the tensile strength.
7. **Stuffing:** During the stuffing process, supplementary materials such as waxes or resins may be incorporated to further improve the leather's characteristics, including its texture or durability.

8. **Setting out:** The setting out or wringing process is performed to eliminate excess tan liquor or moisture by passing the hide through two large rollers.
9. **Pasting:** Pasting involves applying paste or adhesive to leather segments, often serving as a preparatory measure for joining leather components or attaching other materials.
10. **Conditioning:** Conditioning consists of treating the leather with oils, waxes, or other conditioning agents to hydrate and soften it, thereby enhancing its flexibility and durability.
11. **Staking:** Staking is a mechanical technique in leather processing that softens and increases the pliability of the leather by repeatedly stretching and manipulating it with blunt or rounded tools on a staking machine.
12. **Buffing:** Buffing is a mechanical operation where the leather's surface is polished using rotating brushes or abrasive substances. This process can eliminate imperfections, improve texture, and prepare the leather for finishing. This procedure yields a more uniform and smoother surface; however, it compromises the inherent texture and uniqueness present in superior-quality leathers. Typically, corrected grain leather is less expensive and is frequently utilized in mass-produced products such as furniture, automotive interiors, and fashion items. Buffed leather is also referred to as corrected grain. Conversely, full grain leather denotes leather that preserves the complete grain layer, encompassing all of its inherent markings and flaws. This category of leather is regarded as the highest quality due to its exceptional durability and the development of a rich patina as it ages. Full grain leather is commonly utilized in high-end products such as luxury handbags, upscale footwear, and premium furniture.
13. **Finishing:** Finishing represents the concluding phase of leather processing, where the leather's surface is treated to achieve specific attributes such as colour, texture, and sheen. This may involve dyeing, embossing, applying protective coatings, and polishing to enhance both the appearance and durability of the leather.
14. **Glazing:** The glazing process is performed on chromic leather following its seasoning and drying. A glass cylinder, secured to the end of a moving arm, rolls over the leather. The friction generates heat, which softens the wax. This procedure results in a smooth finish and a continuous plastic coating that imparts a high gloss to the grain surface.

15. **Grading:** Grading refers to the evaluation of leather quality based on several criteria, including thickness, texture, colour consistency, and overall appearance. Various grades of leather are designated for different applications, with higher grades generally allocated for premium products.

Following glazing and staking, the leather is smoothly plated on the grain side using a specially designed press. Attractive grain patterns are created through boarding. Vegetable-tanned leathers receive decorative patterns by being embossed with a hydraulic press. The disposal of chrome liquid must be conducted with great care due to its toxic nature. As the leather is prepared, it undergoes grading for temper, uniformity of thickness, colour, and other factors. The graded leather is then appropriately packaged for ease of handling and to achieve a higher market value.

9

Utilization of Poultry Byproducts

The poultry industry generates by-products that encompass all materials derived from farms and processing facilities that are not intended for human consumption. These by-products can be classified as either edible or inedible. The edible by-products primarily consist of tissues and bones extracted from poultry carcasses. Conversely, inedible by-products comprise hatchery waste, which includes infertile eggs, deceased embryos, dead chicks, and eggshells, as well as feathers, blood, offal, fat, and manure. These by-products are typically divided into three categories:

1. By-products from the production phase, which consist of litter and manure sourced from the farm.
2. By-products from hatcheries which include materials originating from the hatchery, such as the shells of hatched eggs, dead embryos, infertile eggs, and deceased chicks.
3. By-products from the poultry processing plant, which encompass blood, feathers, offal (including feet, heads, and intestinal tracts), condemned birds etc.

1. **By-products from the production phase:** Poultry litter and manure from caged layers are extensively utilized as a protein supplement in the diets of both poultry and livestock. Following sterilization and dehydration, poultry manure can be integrated into feed, containing approximately 27-29% protein. It can be incorporated into the diets of broilers at levels of up to 20% and layers at 25%. In addition to their application in animal feed, poultry manure and litter are valuable as a ‘surface dressing’ for agricultural fields. They function not only as effective fertilizers but also improve soil texture. One ton of deep litter yields approximately 29.48 kg of nitrogen (equivalent to 136.08 to 147.42 kg of ammonium sulphate), 20.41 kg of phosphorus (equivalent to 113.40 to 136.08 kg of ordinary superphosphate), and 20.41 kg of potash (equivalent to 45.34 kg of potassium), along with 6.80 kg of magnesium, 6.80 kg of sodium, and 27.21 kg of calcium. Furthermore, poultry manure is rich in trace elements such as boron, copper, iron, sulphur, and zinc.

2. **By-products from hatcheries:** The by-products generated in poultry hatcheries consist of egg shells, infertile eggs, unhatched eggs, and deceased or culled chicks. These components can be transformed into a high-protein meal. The processing of these by-products involves cooking, drying, and grinding, with or without the removal of fat, yielding a meal that contains approximately 18.10% calcium and 413 mg of phosphorus per 100 grams. This meal can be added to layer feed at proportions of 3-5%.
3. **By-products from poultry processing plant:** The dressing of poultry birds results in the generation of various inedible by-products, which include feathers (6-7%), blood (3.5%), heads (3.0%), feet (3.9%), and offals such as intestines, lungs, pancreas, and spleen (8-9% of live weight). These by-products can be processed into valuable end products, including bedding materials, decorative items, sporting equipment, and feather meal. Additionally, they can be converted into poultry by-products meal, hydrolysed feather meal, mixed poultry by-products meal, or poultry grease. The processing entails cooking, drying, pressing to eliminate excess fat, and grinding the material into a fine powder. The final commercial meal should not exceed 16% ash and 4% acid-insoluble ash and can be utilized at levels of 5-7% in feed as a replacement for meat or fish meal.

Table 20: Application of poultry industry waste

By-product	Application
Feather	Livestock feed, as bedding, fertilizer, ornamental use and in sports.
Blood	Livestock feed, as fertilizer, fish bait.
Offal	Livestock feed, pet food, fish food
Mixed poultry by-products meal	Poultry feed (5-7% level)
Hatchery waste Yield 25%, Protein 35% Fat 40% Calcium 0.05% Phosphorus 1%	Poultry feed (3% level)
Poultry manure Moisture 15-18% Protein 25-30% Uric acid 6-8% Ash 15-25% Calcium 3-6% Phosphorus 1.5-2%	Fertilizer In poultry feed e.g. layer ration (up to 15%)

Feather Processing

Feathers constitute approximately 7% of a poultry bird's overall body weight and serve a variety of purposes, including applications in clothing, insulation, bedding, decoration, sports equipment, feather meal, and fertilizer. The characteristics of feathers can vary depending on the species, age, sex of the bird, and the specific location of the feather on the body. Feathers are categorized into several types, including hard feathers, saddle feathers, half fluff, quarter fluff, fluff, plumules, and down.

Down, which serves as the insulating undercoat of lightweight feathers, is characterized by the absence of a shaft or stiff quill and features long fibers. It accounts for about 12-15% of the total feather weight, while the remaining feathers contribute approximately 85-88% of the overall weight.

In the processing of feathers, wing and tail feathers are generally removed prior to scalding, although this practice may not always be observed in commercial operations. Feathers can be preserved for up to 12 hours when immersed in a solution of salt, hydrochloric acid, and water. They undergo multiple washes with mild soap or cleaning agents to eliminate dirt and blood, while ensuring a neutral pH to safeguard the natural oils of the feathers. Dust and dirt may be filtered out as many as 15 times. If feathers are not adequately cleaned, they risk developing mildew or deteriorating. If necessary, bleaching agents such as potassium permanganate, hydrogen peroxide, or chlorine may be employed. Ultimately, feathers are thoroughly rinsed, dried through blowing or steaming to enhance their fluffiness, and sorted by size using air currents.

The use of feathers in bedding has declined in recent years due to the increasing popularity of synthetic fibers and plastic foam; however, high-quality bedding still depends on body feathers, particularly those sourced from waterfowl. For bedding applications, feathers should offer maximum volume during use while minimizing volume for storage. When utilized for decorative purposes, attributes such as colour, shape, size, and plumage patterns become essential. Cock pheasants are especially prized for their vibrant feathers, which are frequently dyed, shaped, and trimmed to achieve specific patterns.

In the realm of sporting equipment, feathers are carefully chosen for their distinct characteristics. For instance, robust feathers sourced from mature turkeys are utilized for the fletching of arrows. To guarantee the correct rotation of arrows, it is essential that all feathers on a single arrow originate from the same wing, whether it be the right or the left. Rigid feathers are employed in the production of shuttlecocks for badminton, while other specially selected feathers are fashioned into artificial lures for fishing.

Among these various uses, feather meal holds the most significant market share. Feathers are predominantly made up of keratin, a complex protein that requires hydrolysis during the rendering process to improve its digestibility for incorporation into animal feed. The flow diagram illustrating the preparation of feather meal is presented in the figure below.

Collection of feathers from poultry processing plant

⇩

Washing with mild soap and water

⇩

Draining and pressing

⇩

Wet rendering at 30-45 psi for one hour

⇩

Cooling and fat removing

⇩

Drying and pulverizing

⇩

Packing in bags

Figure 34: Flow diagram for preparation of feather meal

Table 21: Percent yield and composition of poultry by-products meal

Constituent	Feather meal	Blood meal	Offal meal	Mixed by- products meal
Yield	33	18	55	33
Moisture	7	9	10	8
Crude protein	85	86	52	66
Crude fat	3	1	24	18
Ash	4	3	14	1.8

Feather meal is rich in cystine, threonine, and arginine; however, it is deficient in lysine, methionine, histidine, and tryptophan. To remedy these shortcomings, it is essential to supplement these amino acids in the feed when incorporating feather meal for mono-gastric animals such as poultry and swine. The recommended inclusion level of feather meal in animal diets typically ranges from 0.5% to 1.5%. It is utilized more efficiently by ruminants like cattle, sheep, and goats. The digestibility of feather meal is affected by the duration and pressure of cooking during hydrolysis, with more rigorous processing enhancing the availability of amino acids and their biological value.

Mixed poultry by-product meal, which consists of blood, offal, and feathers in their natural proportions, undergoes rendering and drying. Excess fat may be eliminated, and offals (excluding feathers) are subjected to pressure cooking for approximately 30 minutes prior to drying, cooling, fat extraction, grinding, and packaging. Feathers are hydrolysed separately before being incorporated into the processed offals.

Processing of Hatchery Waste, Inedible Eggs and Egg Shells

a) **Hatchery waste:** Hatchery waste encompasses infertile eggs, deceased embryos, dead chicks or poults, and shells from hatched eggs. In hatcheries that produce egg-laying chickens, this waste may also consist of male chicks that are culled during the sexing process. The contents of the eggs are generally collected, cooked at 10 lbs of pressure for 15 minutes, and subsequently dried. The average yield of dried hatchery by-product meal is approximately 25%. Processing whole eggs with shells presents certain challenges: the calcium-to-other-ingredients ratio becomes excessively high, and the hard, gritty texture of the shells can render the meal less palatable.

b) **Inedible eggs:** Inedible eggs, including those affected by black rot, white rot, sourness, mould, mustiness, decomposition, or physical damage, are initially subjected to cooking, followed by drying and grinding. The resultant processed inedible eggs are commonly utilized as feed for pigs. Nevertheless, given that eggshells constitute 11% of the overall weight of an egg, the use of whole eggs may lead to an excessive calcium concentration, potentially disrupting the nutritional balance of pig rations.

c) **Egg shells:** Eggshells, which account for approximately 11% of the total weight of an egg, are produced in large volumes at egg-breaking facilities and commercial hatcheries. Once collected, the eggshells undergo a drying process and are subsequently heated to 80°C for sterilization, after which they are ground into fine particles to produce eggshell meal.

10

Utilization and Disposal of Slaughter House Organic Waste

In the realm of meat processing, waste is defined as the materials produced during the processing of meat and its byproducts that currently lack economic value and often lead to disposal expenses. This definition may vary depending on market dynamics. The waste generated in meat processing primarily consists of both solid and liquid elements. The solid waste includes manure and residual animal tissues such as meat and fat, while the liquid waste encompasses blood and other bodily fluids. Generally, small remnants of tissue and manure are combined with water and directed to sewage systems. In larger processing facilities, blood is collected and dehydrated for use as animal feed.

It is essential to effectively utilize animal farm residues, which include dung, droppings, and urine. Likewise, innovative approaches should be employed to make use of slaughterhouse waste, which consists of ruminal contents, blood, urine, and trimmings of meat and fat, in order to enhance their value. It is vital to discover comprehensive and inventive methods for utilizing all inedible animal parts. The proper management of slaughterhouse waste not only mitigates pollution but can also generate fuel for lighting and heating within the abattoir itself.

In rural and suburban areas of India, buffalo and cow dung are frequently dried into cakes and utilized as cooking fuel. Furthermore, dung compost contributes to the enhancement of soil humus and fertility, with approximately two-thirds of dung being used as fuel and one-third as fertilizer. Cow dung, which is abundant in organic matter and nitrogen, is also traditionally employed as a floor plaster. In certain regions, sheep owners receive compensation for allowing their flocks to graze on fields post-harvest, thereby improving soil fertility through their manure. Consequently, the organic waste produced by animal industries can be effectively harnessed in numerous ways.

Sources of Wastes from Abattoirs and Meat Processing Plants

The operations involved in slaughtering and processing yield a diverse array of wastes, comprising both solid and liquid components. Given that the raw materials utilized in the meat industry are biological in nature, the resulting waste is rich in organic matter. Significant sources of waste from meat and by-product processing include:

1. **Stockyard waste:** After transport and prior to slaughter, animals are allowed to rest in the lairage, resulting in waste primarily composed of their faeces, urine, and occasionally soil or grit. The quantity and nature of this waste are influenced by various factors, including the species of the animal, its size, dietary habits, the duration of lairage, and the time of feed withdrawal. These wastes are rich in nitrogen, with urine being a major contributor to organic nitrogen, which complicates the treatment of waste. To address this issue, substantial volumes of water are utilized to flush the waste into the wastewater stream, where solids are subsequently separated through sedimentation or screening. Alternatively, urine may be drained away while faeces are permitted to accumulate and dry beneath the floor grating before collection. Although the dry recovery of faeces is preferred, it is not always practical.
2. **Blood collection and processing:** Following slaughter, blood is generally collected for either edible applications, employing hygienic techniques, or for inedible processing. Typically, blood is drained into a collection pit or trough for non-edible purposes. Blood also gathers during the removal of hides or skins, head detachment, and cutting, where it drips and disperses thinly across extensive areas, necessitating regular washing. Due to its rapid coagulation, blood can be gathered from the floor beneath the carcass through dry cleaning methods or by scooping it into a holding bin for subsequent processing with other collected blood. Furthermore, a considerable volume of blood enters the wastewater stream from floor washing in the carcass chiller and during the initial washing and cooling of hides and skins.

   Blood is characterized by high concentrations of total solids, nitrogen, and oxygen demand, rendering it a notable pollutant. Even a minor spillage of blood into wastewater can elevate the pollutant load and treatment expenses. For example, recovering merely 2 litres of blood from a beef carcass can enhance blood yield by 15%, thereby decreasing the Chemical Oxygen Demand of the wastewater by 600 grams and the nitrogen load by 60 grams.

   During the transformation of blood into consumable or non-consumable products such as blood meal, a portion of the blood is frequently lost to wastewater. Generally, blood undergoes coagulation via steam injection, followed by centrifugation and drying to yield blood meal.

This procedure may lead to the loss of blood solids during the stages of coagulation, separation, drying, and the cleaning of blood storage tanks. Consequently, a facility that processes blood on-site must meticulously manage and mitigate these losses to lessen wastewater contamination and enhance overall operational efficiency.

3. **Trimming, cutting, and deboning:** The processes of trimming, cutting, and deboning produce waste that includes meat, fat trimmings, and fine debris from carcass saws. To maximize waste recovery for rendering, it is essential to collect solids through dry cleaning techniques. Small particles that are washed into the wastewater cannot be retrieved through screening or sedimentation and can significantly increase pollutant levels. To avert this issue, floor drains should be equipped with gratings and perforated baskets to prevent larger particles from entering the wastewater.

4. **Viscera processing:** Following the slaughtering and bleeding of animals, evisceration occurs. During this process, gut contents and occasionally gut tissue are introduced into the wastewater stream.

   a) **Paunch emptying:** During the dressing process, the paunch, or stomach section of the animal, is extracted and emptied prior to being sent for rendering or preparation for pet food. The volume and composition of paunch contents differ depending on the species, size, diet of the animal, and the length of time without water and feed before slaughter. For example, a cattle paunch may contain over 80 kg of material. The contents can be expelled using either the wet or dry dumping technique:

   i. **Wet dumping:** The paunches are cut open, and their contents are washed out with water. This method utilizes water to cleanse the paunch and transport the contents out of the facility. However, wet dumping can lead to elevated pollutant levels in the wastewater, as screening and sedimentation can only recover approximately 40% of the total solids.

   ii. **Dry dumping:** The contents are extracted by opening the sac without the use of water, leaving approximately 10% of the contents remaining, which can be cleaned with water if required. Dry dumping typically leads to a reduced pollutant load in the wastewater, as the paunch content is less diluted. This method is preferred for minimizing the pollutant load in the wastewater stream, particularly when the paunch contents are compressed to eliminate excess liquid.

b) **Viscera cutting and washing for rendering:** Inedible gut materials and condemned paunch are processed in the rendering section, where they undergo mechanical crushing by a gut cutter and are subsequently washed using a rotating screen. This procedure produces a significant volume of wastewater with elevated pollutant loads from gut contents, including fat that is dislodged during the cutting and washing processes. Although some solids can be retrieved from the wastewater through screening, the presence of fat complicates their treatment or disposal. Alternatively, directly rendering the gut contents circumvents this issue but is economically impractical due to the deterioration of tallow quality.

c) **Animal casing processing:** In the preparation of casings, faecal material is extracted from the intestine, followed by the cleaning of the mucosa and other tissues from the intestinal wall. This procedure consumes a substantial amount of water, resulting in wastewater with high pollutant loads from the intestinal contents and tissues. Ideally, the waste streams from gut contents and tissues should be segregated to allow for the separate rendering of animal tissues from the gut contents.

5. **Rendering:** The primary sources of waste from rendering are outlined below.

a) **Raw material transport and storage:** Fluids often accumulate from materials that are transported to and stored at a rendering facility, containing blood, fat, and other tissues, which increases the pollutant load in wastewater. To mitigate this, processors can implement dry transport systems, minimize the addition of water to the materials, and apply less pressure to reduce fluid release.

b) **Condensate from cookers and dryers:** The rendering, cooking, and drying processes generate a significant quantity of volatile organic compounds. As the gas streams cool, these compounds condense, aiding in odour control. This condensate introduces a substantial load of nitrogen and organic matter into the wastewater. The volume of condensate produced is affected by the quantity of raw material processed and the water utilized in rendering.

c) **Stickwater:** In the wet rendering process, the aqueous phase that is separated from the meal and tallow is referred to as stickwater. It is isolated through techniques such as decanting, pressing, centrifugation, and drying. Stickwater serves as a major source of pollutants, containing fats, proteins, and minerals. Processors can recover solids from stickwater via evaporation or other physico-chemical methods and subsequently reintegrate these solids into the meal.

d) **Tallow refining:** Following the separation from the meal and aqueous phase, tallow is subjected to a refining process that includes washing and centrifugation to eliminate impurities. This procedure utilizes water, resulting in wastewater that has a high organic load.

6. **Hide and skin processing:** After being removed from the carcass, hides and skins are washed to eliminate blood and loose materials, and then preserved with salt and other chemicals for subsequent processing. During the fleshing stage, flesh and fat tissues are extracted. These operations introduce salt, chemicals, adipose tissues, epidermal tissues, and hydrolysed wool and hair into the wastewater, thereby increasing its pollutant load. As a result, the wastewater exhibits elevated concentrations of chemical oxygen demand and nitrogen.

Utilization of Wastes Using Scientific Methods

Organic wastes generated from the slaughterhouse and meat processing facility can be utilized in various ways:

1. **The generation of biogas for heating and lighting purposes:** Anaerobic digestion is a well-recognized method for harnessing energy from waste through the production of biogas. This technique transforms organic materials such as faeces, urine, and waste from slaughterhouses into biogas (comprising methane and other gases) along with high-quality manure. Typically, biogas consists of 60% to 70% methane, 30% to 40% carbon dioxide, and trace amounts of other gases including ammonia, hydrogen sulphide, and mercaptans. It is also saturated with water vapor and is often referred to as swamp gas, sewer gas, or fuel gas.

The process depends on anaerobic bacteria that flourish in environments devoid of oxygen. For optimal methane production, it is essential to maintain a consistent temperature (generally around 95°F), a stable pH, and a continuous supply of fresh organic material. A decrease in temperature by 20°F can lead to a significant reduction in gas production, potentially by 50%, or may extend the required time for gas production to twice as long. Any deviation from the ideal temperature range can impede or completely stop the digestion process. Anaerobic digestion occurs in two primary stages. Initially, complex organic matter, such as manure, is broken down into simpler organic compounds by acid-forming bacteria. Subsequently, methane-forming microorganisms convert these acids into methane and carbon dioxide.

a) The breakdown of organic compounds into biogas can only occur if the following conditions are met:

i. The raw material must contain nitrogen.

ii. Oxygen must be excluded from the process.

iii. The temperature must be favourable (95°F).

iv. The reaction must be slightly alkaline (with a pH of approximately 7.5).

b) The total production of biogas is influenced by several factors:

i. The type of organic material being digested.

ii. The loading rate of the digester.

iii. The environmental conditions within the digester.

Under optimal conditions (95°F temperature and pH of 7.5), one day's worth of manure from a 500kg cow or buffalo can yield 45 cubic feet of gas. A biogas production facility, regardless of its size, consists of three main components:

i. **Digester:** A tank where fermentation occurs, leading to gas production.

ii. **Gasometer:** A tank for storing gas.

iii. **Pipes:** Utilized for the distribution of gas to designated locations.

Following considerations should be taken into account for the construction of a digester:

- The size of the digester must be sufficient to meet the required output.
- The materials employed in its construction should withstand the pressure exerted by the fermenting and packed materials.
- The digester must be impermeable to both gas and liquid.
- It should be designed for ease of handling and maintenance.
- It must exhibit resistance to corrosion.

A digester intended for anaerobic digestion can take the form of either a cylindrical or square structure and is commonly built from bricks, stone, or concrete blocks. In smaller facilities, a single tank often functions as both the digester and gasometer. The digester is equipped with a water-sealed, movable cover that also serves as a gas storage unit. This cover, or gasometer, may consist of an inverted, movable iron lid that adjusts its position based on the volume of gas it holds. The entire system must ensure both water and gas tightness. The digester receives inputs such as animal blood, urine, dung, ruminal contents, and other effluents at regular intervals. Although this configuration is straightforward and economical, production ceases during the recharging phase of the digester.

In larger facilities, the gas production system generally comprises multiple digesters, where gas is produced and subsequently stored in a separate gasometer. This arrangement facilitates a continuous gas supply, even when one or more digesters are undergoing recharging. When loading a digester with waste material, the water level should be maintained approximately 0.5 meters above the waste. Incorporating some aged sludge can expedite the fermentation process. Gas production typically commences after about a week. A discharge hole located near the bottom of the digester is utilized for the periodic extraction of digested or spent slurry, which can be repurposed as high-quality compost manure.

The dimensions of the digester or tank are influenced by various factors, including the number, size, and species of animals, the volume of dilution water introduced, and the required detention time. A minimum detention time of ten days is mandated; however, extending this duration can enhance the decomposition of waste. Increased detention times require larger tanks. Even after processing, the material may retain an odour and contain nitrogen, phosphorus, and potassium, rendering it significantly polluted. Consequently, the waste should not be discharged directly into water bodies. The volume of the effluent may surpass the initial manure volume due to the addition of dilution water. Lagoons are frequently utilized to store the waste until it can be safely transported for land application.

The biogas generated can enable a meat processing facility to achieve self-sufficiency in light, heat, and energy, and it can also be utilized for domestic applications. The establishment of a biogas plant aids in addressing the disposal issues associated with slaughterhouse waste and holds promise for enhancing rural industries. Nevertheless, methane, which constitutes 6% to 15% of the biogas, has the potential to create an explosive mixture with air. Being lighter than air, it can accumulate in rooftops and confined spaces, complicating detection due to its odourless nature. Methane combusts with a blue flame and produces minimal smoke, with occasional traces of an orange flame. Therefore, it is imperative to implement stringent precautions when processing and handling this gas.

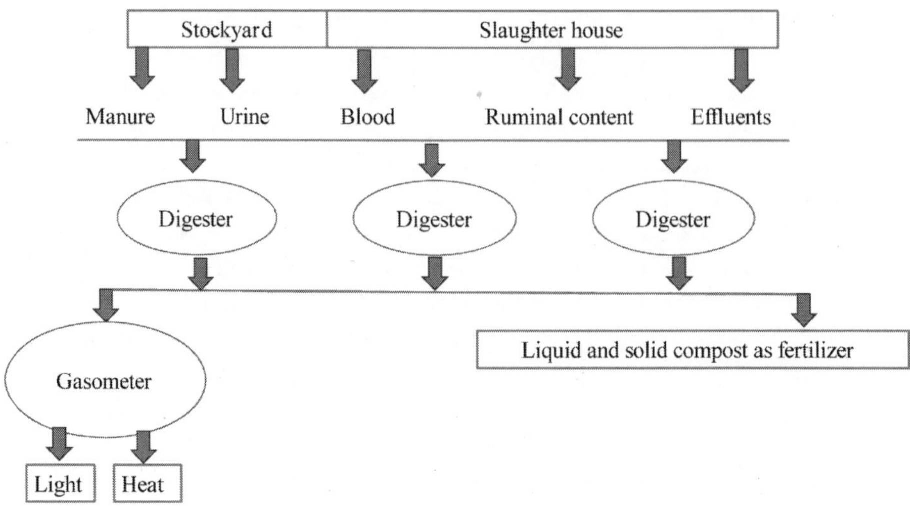

Figure 35: Bio-gas production and utilization

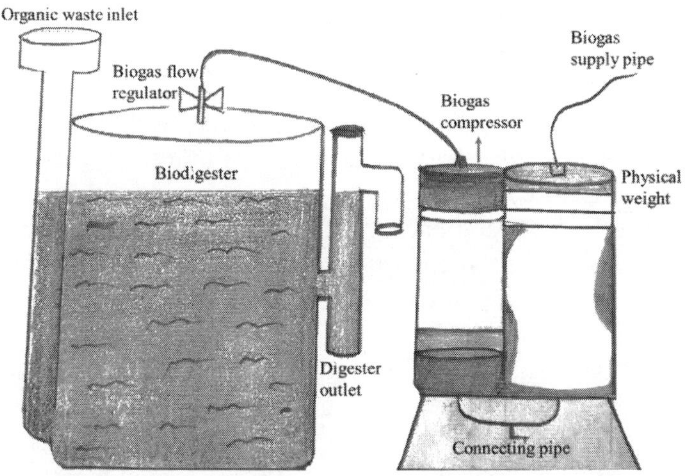

Figure 36: Biogas unit

2. **Production of compost manure:** In regions where the optimal utilization of animal-derived organic matter is not practical, composting emerges as a viable method for generating high-quality manure. This manure supplies vital nutrients, including nitrogen, phosphorus, potassium, and trace minerals, which are crucial for promoting plant growth and enhancing soil structure. Composting facilities are capable of processing a variety of slaughterhouse by-products, such as blood, ruminal and intestinal contents, bedding waste from lairages, meat and

fat trimmings, floor washings, hair, and feathers. Even condemned meat and offal can be incorporated into the composting process when they are chopped into smaller pieces and mixed into the compost pile. The following techniques are typically employed for compost production:

a) **Pit method:** Unused slaughterhouse offal is placed into shallow pits in the ground for composting.

b) **Stack method:** Materials are assembled into a mound above ground, rather than on a concrete surface. Alternating layers of ruminal and intestinal contents, vegetable matter, and slaughterhouse waste are piled up to a height of 2 meters. Coarse materials, such as maize stalks or twigs, are positioned at the base to facilitate ventilation. The mound is then covered with soil or grass for protection. In tropical regions, temperatures can soar to 75°C, aiding in pest control. The mound requires periodic watering and ventilation, often achieved with discarded corrugated sheets. To promote even decomposition, the mound is turned three times at intervals of 20, 40, and 60 days. Typically, composting is completed within approximately 90 days.

c) **Compost bunkers:** Compost bunkers are employed to form clean, orderly piles of compost. They can be built from bricks with adequate openings in the walls. The dimensions of the bunker are determined by the volume of raw materials. The floor should be made of earth, and the surface should be coated with mud. In arid regions, the bunker should be shielded with covering materials such as corrugated iron. The remainder of the composting process follows the same principles as the stack method.

3. **Production of adhesive:** Trimmings, fleshings, pig skin, sinews, tendons, scattered bones, and various animal parts such as horns, hooves, ears, lips, snouts, and tails can be utilized in the production of adhesive. This procedure necessitates only basic equipment and minimal training. Adhesive is extensively employed in sectors including paper, rugs, plywood, carpet sizing, and the manufacturing of synthetic leather.

4. **Production of feed from rumen and intestinal contents:** The undigested material found in the first stomach of herbivores, referred to as ruminal or paunch content, has an approximate weight of 27 kg in cattle, 2.7 kg in sheep, and 1.7 kg in lambs. Within the rumen, fodder is combined with enzymes and decomposed into simpler forms, which are subsequently utilized by rumen microbes. These microorganisms transform the material into high-quality protein and synthesize vitamins, particularly the Vitamin B complex. Consequently, the undigested food

not only preserves the nutritional value of the ingested grass but also becomes enriched with proteins and vitamins.

The liquid and fibrous components of ruminal contents can be separated through pressing. The fibrous component can serve as fuel, while the protein-rich component can be concentrated for use as feed for poultry or pigs. To process the contents, they are dried on concrete platforms, trays, mats, or corrugated iron sheets under direct sunlight. It is essential to maintain a thin drying layer to avoid fermentation of the lower layers. After drying, the material is ground into powder using a mill or pestle and mortar. Occasionally, blood is incorporated with the ruminal contents to improve the protein quality for animal feed. The use of paunch and intestinal contents as feed can lead to a reduction in disposal costs by approximately 50%.

11

Introduction To Wool Fur, Pelt and Specialty Fibers

In textile manufacturing, fibers serve as the fundamental components of raw materials, distinguished by their length, flexibility, and strength, which enable them to be spun into yarns and woven into fabrics. Extremely long fibers are referred to as filaments. Fibers can be classified as natural, derived from plants, animals, or minerals, or artificial, produced through chemical processes. Natural fibers are those sourced directly from animal, vegetable, or mineral origins and can be utilized for creating nonwoven fabrics or spun into yarns for woven textiles. Typically, they are significantly longer than they are wide. Although nature offers a variety of fibrous materials, including cotton, wood, grains, and straw, only a select few are appropriate for textile applications. The appropriateness of a fiber is determined by its length, strength, flexibility, elasticity, abrasion resistance, absorbency, and surface characteristics. Most textile fibers are slender, flexible, robust, and elastic, allowing them to stretch and revert to their original form.

During the 18th and 19th centuries, the Industrial Revolution brought about significant advancements in machinery for processing natural fibers, resulting in increased fiber production. This era saw the introduction of regenerated cellulosic fibers such as rayon, which is produced by dissolving and purifying cellulose, followed by the emergence of synthetic fibers like nylon, which began to compete with the supremacy of natural fibers in the textile industry. Synthetic fibers, with their distinct advantages, started to supplant natural fibers across various sectors. In reaction to this shift, there was a surge in research aimed at enhancing natural fibers through improved strains, advanced production techniques, and modified properties of yarns and fabrics. Nevertheless, synthetic fibers, being more economical and less labour-intensive, captured a larger share of the market.

Natural Fibers: Natural fibers are classified based on their origin:

1. **Vegetable fibers (Cellulose-based):** Vegetable fibers primarily consist of cellulose, but they also include other components such as

hemicellulose, lignin, pectin, and waxes that need to be eliminated during processing.

a) **Seed fibres:** Cotton, kapok, and coir are fibers obtained from the hairs on seeds or the inner surfaces of fruits, consisting of elongated, singular cells.

b) **Bast fibres:** Flax, hemp, jute, and ramie originate from the inner bast tissue of plant stems and are characterized by their overlapping cellular structure.

c) **Leaf fibres:** Abaca, henequen, and sisal are fibers derived from the fibrovascular system of plant leaves.

2. **Animal fibres (Protein-based):** Wool, mohair, and silk are protein-based fibers obtained from animal hair or fur. Silk is particularly unique as it is produced by moth larvae for the purpose of cocoon formation.
3. **Mineral fibres:** Asbestos serves as a prominent example.

Properties of Natural Fibers

1. **Water affinity:** These fibers have the capacity to absorb water, which facilitates the dyeing process.
2. **Non-thermoplastic:** They do not soften when exposed to heat and are not affected by dry heat until decomposition occurs. Additionally, they do not become brittle in cold conditions.
3. **Degradation:** In contrast to synthetic fibers, natural fibers maintain stability under dry heat and do not soften, shrink, or excessively stretch when subjected to heat. They also do not become brittle in freezing conditions. However, they are prone to yellowing due to exposure to sunlight and moisture, and may experience a reduction in strength with prolonged exposure. Natural fibers are vulnerable to microbial decay, such as mildew and rot, especially in environments with high humidity and heat. Cellulosic fibers, including cotton and linen, are particularly susceptible to bacterial and fungal degradation, while wool and silk can also be affected by bacterial and mould damage. Insects like moths, carpet beetles, termites, and silverfish can inflict damage on these fibers. Chemical treatments can enhance the protection of natural fibers against microbial and insect damage, thereby improving their durability.

Specialty Fibers: Specialty hair fibers, esteemed for their fine diameter, natural sheen, and texture, are sourced from particular animals within the goat and camel families. These include:

a) **Mohair:** Sourced from the Angora goat.

b) **Cashmere:** Sourced from the Kashmir goat, often referred to as cashmere wool.

c) **Camel hair:** Primarily sourced from the Bactrian camel.

d) **Llama, Alpaca, Vicuna, and Guanaco:** Llama fiber is thicker and less soft than alpaca fiber, yet it remains prized for its durability and strength. Alpaca fiber is remarkably soft, warm, and lightweight, available in a variety of natural colours. Vicuna fiber is recognized as the finest globally, celebrated for its luxurious quality and high price. Guanaco fiber is coarser than that of alpaca and vicuna, but it is still valued for its strength and durability.

e) **Rabbit hair:** Rabbit hair is silky and prized for its softness and lustre, utilized in high-quality fabrics and yarns.

These fibers are gathered through hunting, domestication, or harvesting fleece from live animals. They generally exhibit less crimp and a lower tendency to felt compared to sheep wool.

Wool: In the early stages of human civilization, sheepskins were primarily utilized for clothing and protection. Over time, the realization emerged that wool could be spun into yarn and subsequently woven into fabric, which signified the inception of woollen textiles. By the era of ancient civilizations, including the Egyptians, Greeks, and Romans, wool had established itself as a fundamental material. These societies refined spinning and weaving techniques, leading to the creation of more advanced fabrics. During the medieval period in Europe, wool became a significant economic force, with a flourishing wool trade, particularly in England and Flanders, which contributed to the expansion of textile industries.

Throughout the ages, selective breeding has enhanced the quality of wool by minimizing coarse outer hairs while increasing softness and fineness. This progression has led to a diverse array of wool types, ranging from fine merino to coarser varieties. The inherent breathability and moisture-wicking characteristics of wool render it suitable for a variety of clothing items, including high-end suits, sweaters, and everyday wear. Additionally, it is preferred for carpets and rugs due to its durability and enduring aesthetic. As a renewable and biodegradable resource, wool presents a sustainable alternative to synthetic fibers. The industry has increasingly focused on eco-friendly practices and has innovated performance wool that is lightweight, moisture-wicking, and odour-resistant, making it ideal for activewear. Specialty wools such as cashmere and alpaca, renowned for their remarkable softness and warmth, are utilized in luxury garments and premium textiles.

Wool is primarily obtained through the shearing of fleece from live sheep. Wool derived from slaughtered sheep, referred to as pulled wool, is considered to be of lower quality. Following shearing, the fleece undergoes cleaning to eliminate "wool grease," a fatty substance that is processed into lanolin, which is used in cosmetics and ointments.

Wool fibers consist of keratin, a protein that is more susceptible to chemical damage and environmental influences compared to plant-based fibers. The diameter of wool fibers typically ranges from 16 to 40 microns. Fine wools measure approximately 1.5 to 3 inches (4 to 7.5 cm) in length, whereas coarser fibers can reach lengths of up to 14 inches (35 cm). Fine fibers exhibit up to 30 waves per inch (12 per cm), while coarser fibers display a reduced number of waves. Generally, wool appears whitish, although it can also be found in brown or black shades, with coarser wools exhibiting a higher lustre.

Individual wool fibers can support weights ranging from 0.5 to 1 ounce (15 to 30 grams) and can stretch between 25 to 30 percent of their length without breaking, although they exhibit reduced strength when wet. Wool fibers possess the ability to revert to their original length after being stretched or compressed, which contributes to the fabric's ability to maintain its shape, drape effectively, and resist wrinkling. The natural crimp present in wool fibers facilitates interlocking, resulting in the creation of strong and insulating yarns and fabrics. Wool can absorb moisture equivalent to 16 to 18 percent of its weight, which aids in temperature regulation and warmth, and it dries slowly without causing rapid chilling to the wearer.

Wool demonstrates excellent dye affinity, enabling it to absorb vibrant colours. It is recommended to wash wool with mild detergents at temperatures below 20°C (68°F) and with minimal agitation. While wool is resistant to dry-cleaning solvents, it is sensitive to strong alkalis and high temperatures. Contemporary treatments have enhanced wool by providing resistance to insects and mildew, controlling shrinkage, improving fire resistance, and offering water repellence. The inherent properties of wool (durability, resilience, and comfort) render it a versatile and valuable fiber. Innovations in processing and finishing continue to enhance its functionality and significance within the textile industry.

Woolen yarns, composed of shorter fibers, are thick and substantial, making them perfect for tweed fabrics and blankets. Worsteds, crafted from longer fibers, are finer, smoother, and more resilient, rendering them appropriate for high-quality dress fabrics and suits. In the United States, wool that has not been previously utilized is referred to as new or virgin wool. Given the restricted global supply, recovered wools are frequently employed; in the U.S., wool derived from unused fabric is termed reprocessed wool, while wool sourced from used materials is identified as reused wool. Recovered wools, which are commonly utilized in woollens and blends, are generally of lower quality due to damage incurred during recovery. Australia, Russia, New Zealand, and Kazakhstan are leaders in the production of fine wool, whereas India stands as the foremost producer of coarser carpet wools. The primary consumers of these products include the UK, the US, and Japan.

Mohair: Mohair is a luxurious fiber obtained from the Angora goat, with its name tracing back to the Arabic term "mukhayyar," which translates to "goat's hair fabric," evolving into "mockaire" in early European contexts. As one of the oldest textile fibers, mohair has been cultivated in Turkey for millennia and gained prominence in European textile industries during the 19th century. In the mid-1800s, Angora goats were introduced to southern Africa and the southwestern United States to improve the quality of local goat herds.

Production and processing: Angora goats yield fleece in uniform locks, growing annually at a rate of 8 to 12 inches (20 to 30 cm). They are sheared biannually, with each shearing producing approximately 5 pounds (2.25 kg) of fleece. Contemporary breeding techniques have reduced the amount of coarse outer guard hair, emphasizing the softer inner fiber. In the U.S., mohair fleece is typically marketed through local warehouses or sold directly to mills, with significant markets located in Boston and Philadelphia. Turkish mohair is mainly traded in Istanbul, while African mohair is predominantly exported to the UK. The fleece undergoes processing to eliminate natural grease, dirt, and vegetable matter, resulting in a cleaned fleece yield of 70 to 90 percent of the original weight.

Fiber characteristics: Similar to wool, mohair primarily consists of keratin; however, it has a distinct structure with a reduced number of scales on its outer layer, leading to a smoother texture. Mohair is characterized by its length, lustre, strength, resilience, and durability, along with its ability to absorb moisture and affinity for dyes. Nevertheless, it exhibits greater sensitivity to chemicals. In terms of its response to heat, sunlight, and aging, mohair behaves similarly to wool, yet it is less prone to felting due to its smoother scale structure.

Uses: Fabrics made from mohair, often in a pile format, are utilized in a wide range of garments, including outerwear, summer suits, and dresses, as well as in knitted products and yarns. It is frequently blended with other fibers or employed in warp or filling yarns within woven textiles. Although it was once favoured for upholstery, the popularity of mohair has diminished with the advent of synthetic fibers. The primary producers of mohair are Turkey, the United States, and southern Africa, while the main consumers include the UK, Netherlands, and Belgium, with demand fluctuating based on fashion trends.

Cashmere: Cashmere is a luxurious, downy fiber sourced from the undercoat of the Kashmir goat, setting it apart from other soft wools. In certain Asian regions, it is referred to as pashmina and has gained fame for its exquisite shawls originating from Kashmir, India. Authentic cashmere is derived from the soft undercoat, while the coarser outer fibers measure between 4 to 20 cm in length, and the delicate cashmere fibers range from 2.5 to 9 cm. The majority of cashmere is hand-harvested during the moulting season, although in Iran, shearing is practiced. The fleece undergoes cleaning and processing to eliminate impurities and coarse hairs, which significantly affect its final quality and market price. High-quality cashmere is highly valued for its warmth, softness, and drape, yet it is more vulnerable to abrasion and damage from strong alkalis and elevated temperatures. It is commonly used in premium garments and blends. Major producers of cashmere include China, Mongolia, and Iran, while the leading consumers are the U.S., U.K., and Japan.

Rabbit Hair: Rabbit hair is derived from both Angora and common rabbits. The fibre from Angora rabbits, predominantly produced in France and England, is characterized by its silky texture, white colour, and is highly esteemed for its softness and lustre, making it suitable for high-quality fabrics and yarns. Each Angora rabbit typically produces approximately 200–400 grams (7–14 ounces) of fibre each year. In contrast, common rabbit hair, sourced from domesticated white and grey rabbits, is coarser and is primarily utilized for felt and knitted products. Both varieties are frequently blended with other fibres to enhance warmth and softness. Within the fur industry, rabbit fur is extensively utilized; however, it is delicate and is often dyed and processed to mimic the appearance of more luxurious furs.

Horse Hair: Horsehair is a fibre obtained from the manes and tails of horses, generally measuring between 8 inches (20 cm) and 3 feet (90 cm) in length, and is often black in colour. It is recognized for its coarseness, durability, and shiny appearance, featuring a hollow central canal or medulla that contributes to its relatively low density. Mane hair is softer, with a diameter that ranges from 50 to 150 microns, whereas tail hair is coarser and more robust, with a diameter between 75 and 280 microns, and is typically sold separately.

Uses

- **Fabrics:** The longest strands are utilized in textiles.
- **Brushes:** Medium-length strands are employed for paint, industrial, and household brushes.
- **Stuffing:** Shorter strands are used as stuffing in furniture and mattresses.
- **Musical instruments:** High-quality white horsehair is utilized for premium violin bow strings.

Horsehair fabric, commonly referred to as haircloth, is characterized by a rigid, open weave and integrates lengthwise yarns of another fibre, such as cotton, with crosswise yarns of horsehair. Historically, it was used for interlining in clothing and millinery (women's hats), but has largely been supplanted by synthetic materials. Key producers of horsehair include Argentina and Canada, with additional sources from Mongolia, China, and Australia.

Camel Hair: Camel hair is a natural fiber derived from the fur of camels, particularly the Bactrian camel indigenous to Central Asia. It is highly valued for its softness, warmth, and durability. The fiber is generally harvested during the camel's moulting season, cleaned, and processed to eliminate impurities. Camel hair is utilized in a variety of textiles, such as coats, sweaters, and blankets, owing to its insulating properties and strength. It can be blended with other fibers to enhance functionality and is esteemed in high-end clothing and luxury products.

Fiber characteristics

a) **Softness:** Camel hair is recognized for its soft texture, which enhances comfort when worn.

b) **Warmth:** The fiber offers exceptional insulation, making it suitable for garments designed for cold weather.

c) **Durability:** Camel hair is robust and resilient, contributing to its longevity in textile applications.

d) **Lustre:** The fiber possesses a natural sheen, although it is less lustrous than silk.

e) **Lightweight:** In spite of its warmth, camel hair is relatively lightweight, which adds to its comfort and versatility.

Processing

1. **Harvesting:** Camel hair is gathered during the moulting season when camels lose their winter coat. The hair is usually collected through combing or shearing, with care taken to preserve the quality of the hair.
2. **Cleaning:** Following harvesting, the camel hair undergoes cleaning to remove impurities such as dirt, grease, and plant matter. This cleaning process typically involves washing the hair in water and occasionally using mild detergents.
3. **Sorting:** The cleaned hair is sorted based on its quality. Camel hair can differ in texture, colour, and length, and sorting ensures that fibers with similar characteristics are grouped together.

4. **Carding:** The sorted hair is carded to disentangle and align the fibers. This step employs mechanical brushes to separate the fibers and prepare them for spinning.
5. **Spinning:** The carded camel hair is spun into yarns. The spinning process twists the fibers together to form a continuous strand of yarn that is suitable for weaving or knitting.
6. **Dyeing:** Camel hair yarns can be dyed in a variety of colours. The dyeing procedure resembles that of other natural fibers, which involves immersing the yarn in dye solutions.
7. **Weaving/Knitting:** The dyed yarns are either woven or knitted into fabric. Camel hair can be utilized to produce different types of textiles, including woven materials and knitted clothing.
8. **Finishing:** The completed fabric undergoes finishing processes to improve its texture, appearance, and functionality. This may involve techniques such as softening, brushing, or shaping the fabric.

Uses

1. Apparel

- **Outerwear:** Camel hair is employed in the production of warm and durable outerwear, such as coats and jackets. Its insulating characteristics render it ideal for cold environments.
- **Sweaters and cardigans:** Sweaters and cardigans made from camel hair are appreciated for their softness and warmth.
- **Suits and trousers:** Camel hair is occasionally incorporated into suits and trousers, providing a luxurious sensation and effective thermal insulation.

2. Home furnishings

- **Blankets and throws:** Camel hair blankets and throws are recognized for their warmth and comfort, making them favoured choices in bedding.
- **Upholstery:** The fiber is utilized in premium upholstery fabrics for furniture, contributing a sense of luxury and durability.

3. Accessories

- **Scarves and shawls:** Camel hair is featured in scarves and shawls due to its softness and warmth.
- **Gloves and hats:** The fiber is also employed in the manufacture of gloves and hats, offering insulation and comfort.

4. Specialty uses

- **High-quality rugs:** Camel hair is used in the creation of fine rugs and carpets, particularly in areas where camel hair is traditionally woven into textiles.
- **Traditional garments:** In certain cultures, camel hair is utilized to craft traditional garments and textiles, highlighting its historical and cultural importance.

Camel hair continues to be a highly regarded textile fiber because of its distinctive properties and versatility. Its inherent warmth, softness, and durability make it appropriate for a broad spectrum of uses in both apparel and home furnishings.

Silk: Silk is a natural fiber recognized for its smooth texture, opulent lustre, and durability. It primarily originates from the cocoons of the silkworm *Bombyx mori* and is greatly esteemed in the textile industry.

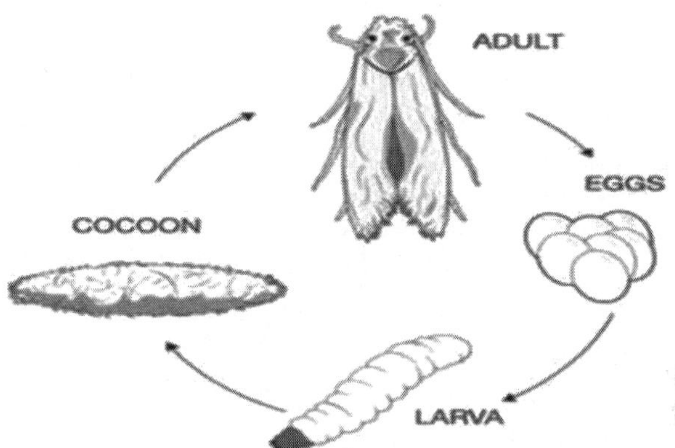

Figure 37: Lifecycle of silkworm

Fiber characteristics

- **Lustre and sheen**: Silk is renowned for its inherent lustre, which imparts a sense of luxury. This brilliance is attributed to the fibre's triangular prism-like configuration, which refracts light.
- **Strength:** In spite of its fragile look, silk is a robust fibre. It possesses high tensile strength, rendering it durable and resistant to wear.
- **Softness:** The fibre exhibits a smooth texture that feels gentle against the skin.
- **Elasticity:** Silk demonstrates good elasticity, allowing it to stretch and revert to its original form, although it is not as elastic as certain synthetic fibres.

- **Absorbency:** It effectively absorbs moisture, contributing to comfort during wear and facilitating dyeing processes.
- **Thermal properties:** Silk has the ability to regulate temperature, keeping the wearer warm in cooler climates and cool in warmer conditions.

Processing

1. **Harvesting:** The primary source of silk is the cocoon of the silkworm. Once the worms have spun their cocoons, they are boiled or steamed to soften the sericin, a protein that binds the silk fibres together.
2. **Reeling:** The softened silk cocoon is unwound in a procedure known as reeling. The continuous filament of silk is meticulously unwound from the cocoon and wound onto reels.
3. **Cleaning:** The raw silk undergoes cleaning to eliminate sericin and other impurities. This process involves washing the silk in a hot solution to dissolve the sericin.
4. **Spinning:** The cleaned silk filaments are spun into threads or yarns. The spinning process may involve twisting and combining multiple filaments to achieve the desired thickness and strength.
5. **Dyeing:** Silk can be dyed through various techniques. Its high absorbency allows it to take on dyes effectively, resulting in rich, vibrant colours.
6. **Weaving/Knitting:** The dyed silk yarns are woven into fabric or knitted. Silk can be crafted into various types of fabric, including satin, chiffon, and crepe.
7. **Finishing:** The final fabric undergoes finishing processes to enhance its texture, appearance, and performance. This may include techniques such as calendaring (to smooth the fabric) or applying a finish to improve water resistance or softness.

Uses

1. Apparel

- **High fashion:** Silk is a favoured material for high-end fashion garments due to its opulent texture and lustre. It is utilized in the creation of dresses, blouses, and suits.
- **Formal wear:** Silk is frequently employed in formal attire such as evening gowns, wedding dresses, and tuxedos.
- **Lingerie:** The gentle and sleek nature of silk renders it a preferred fabric for lingerie and sleepwear.

2. Home furnishings

- **Upholstery:** Silk is utilized in premium upholstery for furniture owing to its resilience and visual appeal.
- **Bedding:** Silk pillowcases and sheets are esteemed for their smooth feel and potential advantages for skin and hair.

3. Accessories

- **Scarves and ties:** Silk is often used for scarves, ties, and shawls because of its capacity to retain vivid colours and intricate patterns.
- **Handkerchiefs:** Silk handkerchiefs are appreciated for their luxurious texture and aesthetic.

4. Specialty uses

- **Embroideries and decorations:** Silk threads are employed in detailed embroidery and decorative arts.
- **Parachutes and medical sutures:** Historically, silk was utilized in parachutes and medical sutures due to its strength and flexibility.

5. Art and culture

- **Traditional textiles:** Silk holds a significant position in traditional textiles and garments across diverse cultures, including Chinese silk robes and Japanese kimonos.

Silk continues to be a highly esteemed fiber in both the fashion and textile sectors, recognized for its distinctive characteristics and luxurious attributes. In spite of the competition posed by synthetic fibers, the artistry and natural elegance of silk maintain a unique status in textiles.

Pelt: The production and processing of pelt involves the treatment of animal hides, which are typically used to manufacture various leather goods.

Production

1. **Harvesting:** Pelts are acquired from animals through the process of skinning. This can take place in slaughterhouses, where pelts serve as a by-product of meat production, as well as from animals that are specifically bred for their pelts, such as mink, fox, or rabbit.
2. **Preservation:** To avert decomposition, pelts are preserved promptly after skinning. Common methods of preservation include:
    - **Salting:** Applying salt to the pelt to extract moisture and inhibit bacterial proliferation.

- **Drying:** Air-drying the salted pelt to further prevent deterioration.
- **Pickling:** Immersing the pelt in a solution of acids and salts to assist in its preservation.

Processing

1. **Soaking:** The dried or pickled pelts are immersed in water to rehydrate them, preparing them for subsequent processing.
2. **Fleshing:** Any residual flesh and fat are eliminated from the pelt using a fleshing machine or knife.
3. **Tanning:** The pelt is subjected to tanning to transform it into durable leather. Common tanning techniques include:
    - **Chrome tanning:** Utilizing chromium salts, resulting in leather that is soft, flexible, and water-resistant.
    - **Vegetable tanning:** Employing tannins from plant sources, yielding a firmer leather with a unique colour and scent.
    - **Aldehyde tanning:** Applying aldehydes, which produces soft and resilient leather.
4. **Dyeing and Finishing:** The tanned pelt is dyed to achieve the desired hue and subsequently finished to enhance its texture and appearance. Finishing processes may encompass oiling, buffing, and the application of protective coatings.

Uses

- **Apparel:** Pelts are utilized to create various types of garments, including coats, jackets, and hats. High-quality pelts, such as those from mink or sable, are especially valued in the fashion sector.
- **Accessories:** Pelts are also employed in the production of accessories like gloves, scarves, and bags.
- **Home furnishings:** They are incorporated into upholstery and decorative items, including rugs and throws.
- **Crafts and specialty items:** Pelt leather is utilized in a range of craft items and specialty products, including certain types of footwear and handmade goods.

Section-II: Wool Processing

12

Glossary of Terms of Wool Processing

Apparel wool: Any wool utilized for the creation of garments, excluding carpet and pulled wool, is referred to as apparel wool.

Bale: A bale is a wool package that contains a designated minimum weight of wool. The wool is compactly packed into bales, entirely covered with new gunny cloth, and secured with multiple iron hoops. Each bale typically weighs between 100 kg and 200 kg, which is considered a standard commercial weight.

Belly wool: Belly wool is the wool that grows on the underside of sheep.

Black wool: Black wool encompasses any wool that is black, brown, or gray in colour.

Blending: The process of combining various grades and/or lengths of wool in either their raw or semi-compressed state to produce a specific type of yarn is known as blending.

Blood grade: Blood grade initially indicated the percentage of Merino (fine wool) genetics present in a fleece.

Break: A break is a noticeable weak point along the locks in fleeces, resulting from a decrease in the diameter of the wool fibers. This condition is usually caused by illness, fever, or significant stress.

Bright wool: Bright wool is grease wool that is almost white and exhibits minimal yellow coloration due to excess yolk.

Britch or breech wool: Britch or breech wool pertains to the wool harvested from the hindquarters of sheep, which is generally the coarsest section of the fleece.

Buck fleeces: Buck fleeces are those obtained from mature rams. They are usually longer and coarser than ewe wool from the same breed and possess a distinct ram odour.

Burr: Burr denotes vegetable matter, such as twigs and straw, that can be found in wool.

Carbonizing: The process of treating wool with chemicals (acids) to eliminate or destroy burr (vegetable matter) while causing minimal damage to the wool fibers is referred to as carbonizing.

Card: A card is a machine employed to separate wool by opening locks or tufts.

Carding: Carding is a phase in yarn production where fibers are brushed and aligned to enhance parallelism. This procedure also removes a significant amount of foreign matter, resulting in a manageable product known as sliver.

Carding wool: Carding wool denotes short-stapled wool that is optimally suited for the production of woollen yarn.

Carroting: Carroting involves the chemical treatment of the tips or ends of rabbit wool to either stiffen or soften them, thereby inducing artificial crimp.

Carpet wool: Carpet wool is characterized as a type of coarse, harsh, and robust wool that is particularly appropriate for carpet manufacturing. It is often sourced from unimproved sheep and is favoured for its durability in carpet production.

Clean wool content: This term refers to the quantity of clean scoured wool that remains after the elimination of all vegetable matter and other foreign substances, ensuring it contains 12% by weight of moisture and 1.5% alcohol-extractable matter.

Clip: The term clip pertains to the annual yield of wool from a specific flock, state, or country.

Clothing wool: This refers to wool fibers that are too short for combing, utilized in the production of woollen yarn; it is commonly known as carding wool.

Coloured wool: This term describes fleeces from sheep that have been specifically bred to produce naturally coloured fibers, primarily for use by handcrafters.

Combing: Combing is an intermediate step in the preparation of worsted yarn that aligns longer fibers in parallel while eliminating short fibers (noil), broken ends, tangled naps, and larger vegetable particles. Wool with fibers measuring 6.5 cm in length is termed combing wool.

Coring: Coring is a technique for sampling wool bales, bags, or fleeces. The cores (samples) are analysed to estimate the clean wool content (yield), fiber diameter, and other factors, including the type and amount of vegetable matter present.

Cotty wool: This refers to wool that has become matted or felted on the sheep's body due to an insufficient amount of grease.

Count: Count is a numerical representation that defines the fineness degree of a fiber.

Crimp: Crimp describes the wavy structure of wool fibers. The number of crimps has an indirect relationship with the fineness of the fibers; a higher number of crimps indicates better quality in terms of fineness.

Crutching: Crutching refers to the procedure of extracting wool from the regions surrounding the dock and/or udder of sheep.

Cotted wool: Cotted wool denotes wool fibers that have become intertwined due to felting. This classification is deemed inferior and typically arises from inadequate grease in the wool or the sheep's poor health.

Delaine wool: Delaine wool is distinguished by its long-stapled, fine fibers.

Dingu: Dingu describes wool that is devoid of brightness or lustre.

Dead wool: Dead wool is derived from deceased sheep, exhibiting variations in shrinkage, softness, and strength when compared to wool sourced from living sheep.

Defective wool: Defective wool encompasses wool that possesses flaws diminishing its value, such as damage caused by fire, water, or moths. Burry wool is frequently categorized as defective.

Denier: This term signifies the fineness of a yarn, measured as the weight in grams of 900 meters of yarn. The relationship between Tex and Denier is expressed as: Tex = (Denier X 0.1111).

Density: This refers to the quantity of fibers that grow on a specific area of a sheep's skin. A higher number of fibers indicates greater density.

Felt: Felt is a textile material produced by matting, compressing, and pressing fibers together. In contrast to woven or knitted fabrics, felt is formed through the entanglement of fibers, often utilizing heat, moisture, and pressure, which leads to the interlocking of fibers and the creation of a dense, non-woven fabric. Felt can be crafted from various materials, including wool, fur, or synthetic fibers, and is recognized for its durability, warmth, and versatility in applications ranging from clothing and accessories to crafts and industrial uses.

Felting: Felting is the characteristic of wool and certain other animal fibers that enables them to closely intertwine and interlock, forming a compact mass when subjected to agitation in warm, moist conditions.

Fine: This term refers to an American grade of wool that was initially applied to fleeces from pure Merino breeding.

Fleece: Fleece pertains to the total wool harvested from a single sheep during one shearing.

Fleece density: Fleece density serves as an index indicating the number of wool fibers present per unit area of a sheep's body.

Fellmongering: The act of extracting wool from the pelt through bacterial processes or chemical treatments is referred to as fellmongering.

Fur: The term fur denotes any animal skin or its parts that retain hair, fleece, or fur fibers, whether in their raw or processed forms, but excludes skins intended for leather production or those from which hair, fleece, or fur fibers have been entirely removed during processing.

Grade: The term grade signifies the classification of wool based primarily on the fineness and length of its fibers.

Grease: Grease encompasses all impurities present in unscoured wool, including yolk, suint, and soluble foreign substances, while excluding vegetable matter.

Hank: A hank is defined as 560 yards of worsted yarn. The quantity of these standard hanks that collectively weigh one pound determines the yarn's count or size.

Harsh wool: Harsh wool is characterized by a lack of softness and a wiry texture, typically sourced from British meat breeds.

Heavy wool: Heavy wool is identified by a significant amount of impurities, particularly sand and dirt, leading to a reduced yield of usable fiber.

Hide: Hide refers to the outer covering of a mature or fully grown animal of larger species, such as cattle, horses, camels, and buffalo.

Kemp: Kemp is defined as coarse, opaque, and highly medullated wool fibers that periodically shed and are regarded as a notable defect. Commonly found around the hindquarters of sheep, kemp fibers possess minimal desirable traits compared to standard wool fibers. They are typically chalky white, lacking lustre, and feature a thick central medulla with a hollow core, rendering them coarse, straight, and non-felting. Although kemp is generally classified as a rejected grade, it is still considered more valuable than ordinary hair.

Keratin: Keratin is a type of protein present in hair, wool, hooves, feathers, and horns.

Knitting: Knitting refers to the technique of fabric or garment creation through the interlacing of yarn loops.

Lamb's wool: Lamb's wool is the fleece obtained from lambs, which is generally finer, shorter, and softer than the wool harvested from the same breeds of adult sheep.

Lanolin: Lanolin is a refined form of wool grease, primarily composed of a blend of cholesterol esters. It is frequently employed in cosmetic products and can also serve as a lubricant for aircraft engines.

Leather: Leather is a broad term that describes hide or skin which maintains its original fibrous structure to a significant extent and has undergone treatment to become resistant to decay, even when exposed to water.

Lustre: Lustre denotes the inherent gloss or sheen present on a fiber, which arises from the reflection of light. This quality is commonly observed in mohair and wool derived from long wool breeds.

Medullated fibre: Wool fibers that possess a hollow structure (having a medulla) as opposed to being solid, as is the case with true wool, are referred to as Medullated fibre.

Metric count: Metric count indicates the length in kilometers of one kilogram of yarn and is represented as Nm.

Noil: Noil pertains to the short fibers that are extracted from wool during the combing process.

Pelt: Pelt signifies the hide or skin that has been prepared for tanning through the removal of hair or wool, epidermis, and flesh. Plain wool: Plain wool is defined by its minimal crimp.

Pulled wool: Pulled wool is the wool that is extracted from the skin of sheep that have been slaughtered.

Roving: A roving is a long, narrow bundle of fibers that has been twisted to keep the fibers together.

Scoured wool: Scoured wool refers to wool that has been cleaned through washing to eliminate grease, dirt, and suint.

Scouring: Scouring is the process of cleaning wool by washing it with alkali and detergent to remove dust, dirt, and soluble substances.

Seedy wool: Seedy wool is characterized by the presence of seeds, typically from grasses or weeds, which must be carbonized for their removal.

Shearing: The process of removing the natural fleece of sheep by hand, chemically, or mechanically is known as shearing.

Skin: Skin refers to the outer covering of smaller animals (such as sheep and goats) or the immature forms of larger species (like calves and ponies).

Skirting: Skirting describes the wool obtained from the lower parts and legs of a sheep's body, which is often dirty and stained.

Sorting: Sorting is the method of categorizing a fleece into various segments or sorts primarily based on its fineness and length.

Sound wool: Sound wool refers to robust wool that is devoid of breaks or tenderness.

Spinning: Spinning is the procedure that converts continuous, untwisted strands of fibers into the desired yarn count and twist, rendering it appropriate for subsequent processing.

Staple: Staple denotes a cluster or group of wool fibers that are naturally aggregated within a fleece.

Staple length: Staple length pertains to the measurement of the wool fibers in a fleece, specifically denoting the longest fibers present in that fleece. This measurement is a crucial element in wool grading and influences the quality and applicability of the wool for various uses.

Suint: Suint is a secretion from the sheep's body, primarily composed of potassium salts of fatty acids along with smaller quantities of sulphates, phosphates, and nitrogenous substances. Suint is responsible for the distinctive odour associated with sheep.

Tex: Tex is a measurement system for indicating fiber diameter based on 1,000 meters of fiber for a yarn count. It is determined by the weight in grams of one kilometer of yarn.

Top: Top is an uninterrupted untwisted strand of wool fibers of a specified length from which shorter fibers have been eliminated.

Twist: Twist refers to the spiral arrangement of yarn components, typically resulting from the relative rotation of two ends.

Virgin wool: Virgin wool is defined as wool fibers that have been directly harvested from sheep (either shorn or pulled) and subsequently processed into yarn and/or fabric.

Weaving: Weaving is the process of fabric or garment creation through the interlacing of threads.

Wool ball: A wool ball is a mass of wool that has become tangled into a ball, typically found in the first or fourth stomach of lambs. Wool is ingested from the mother's fleece during suckling, which can accumulate and sometimes obstruct the stomach outlet, potentially leading to death.

Wool blindness: Wool blindness is a traumatic ocular condition that impacts sheep with heavily wooled faces. The wool can extend over the eyes, obstructing vision and potentially causing damage to the eye tissues. A common solution is to regularly trim the wool surrounding the eyelids.

Wool classing: Wool classing involves the grading of skirted fleece or pieces prior to their packing in shearing sheds, a practice typically observed in New Zealand and Australia.

Wool grease: Wool grease is a complex amalgamation of esters secreted by the sebaceous (wax) glands located in the skin of sheep.

Wool maggots: Wool maggots, also referred to as fleece worms, are primarily linked to the black blow fly, blue bottle fly, and green bottle fly. These flies deposit their eggs on the wool of a sheep's hindquarters and thighs, often in regions contaminated by faeces or urine. Sheep that are sick, injured, or lying down in wet conditions are particularly susceptible to fly infestations. Wounds that bleed into the wool can also attract wool maggots. The eggs hatch shortly after being laid, and the young maggots commence feeding on the filth and the yolk of the wool.

Wool rot: Wool rot is a fungal infection caused by *Dermatophilus dermatonomus*, resulting in skin irritation in sheep during wet conditions and the development of hard, yellowish scabs. Although the skin lesions may eventually heal, the growing wool can carry the scabs away, leading to a condition known as lumpy wool.

Wool sorters/stapler: Wool sorters, or staplers, are individuals tasked with sorting wool according to various criteria such as fibre type, fineness, and length.

Wool sorters disease: Wool sorters' disease is an alternative term for anthrax, a severe infectious disease that can affect individuals who handle contaminated animal products, including wool.

Wool sorting: Wool sorting is the procedure of dividing a fleece into distinct sections based on fineness and length. This is distinct from grading, where fleeces are evaluated as a whole without subdivision.

Woollen fabrics: Woollen fabrics are textiles made from woollen yarn, noted for their warmth and texture.

Yolk: The yolk consists of a combination of various substances, with cholesterol being the primary component that seems to safeguard fibers against the harmful impacts of environmental conditions. Once it is cleaned and refined, yolk produces lanolin, which is utilized in ointments, cosmetics, leather treatments, and as a rust inhibitor.

13

Basic Structure and Development of Wool Follicle

Wool serves as a dense, wavy, and fibrous protective layer for sheep, primarily made up of the insoluble protein keratin. The wool fiber develops from follicles located in the dermis, which is the skin's middle layer. Each wool fiber comprises two distinct layers:

1. **Outer layer:** This layer consists of flat, irregularly shaped scales that overlap, resembling fish scales. These scales are less tightly bound than those found in hair and are essential to the wool's characteristics, such as its capacity to shrink, strengthen, and felt. The outer layer also affects the overall quality of the wool. While the sheep is alive, the outer scales are coated with wool sweat or grease (termed suint), secreted by specialized glands to preserve the fiber's condition. This grease can be partially removed during washing, and its presence influences the wool's condition, which pertains to the quantity of grease or oil present in the wool.
2. **Inner layer:** Known as the cortex, this layer is composed of long fibrils that are bonded together. In contrast to hair, which contains an air-filled medulla in its inner layer, wool does not possess this feature, contributing to its superior strength compared to hair.

Morphological Structure of Wool: Technically, wool comprises three layers: cuticle or epidermis, cortex and medulla.

1. **Cuticle:** The outermost layer of the wool fiber, referred to as the cuticle, is made up of flat, irregularly shaped scales that overlap with their edges directed toward the fiber tip. This overlapping configuration, akin to roof tiles, produces a rough texture that increases friction and aids in felting. When disturbed, these scales interlock, leading to the entanglement of fibers and the formation of a matted fabric, which is beneficial for felted textiles. Additionally, the cuticle possesses a waxy coating that offers a water-repellent barrier, preventing water penetration while permitting moisture vapor absorption, thereby enhancing wool's resistance to water-based stains and improving comfort by wicking moisture away.

2. **Cortex:** The cortex constitutes the core of the wool fibre, comprising elongated, slightly flattened, and twisted spindle-shaped cells. It plays a vital role in the fibre's strength, elasticity, and dyeing characteristics. Approximately 90% of the fibre is made up of the cortex, which includes two distinct types of cells:
    a) **Ortho-cortical cells:** These cells exhibit greater elasticity and hydrophilicity, significantly expanding when they absorb moisture.
    b) **Para-cortical cells:** In contrast, these cells are less elastic and more hydrophobic, showing minimal expansion upon moisture absorption.

    The variation in expansion between these cell types results in the crimping of the wool fibre, a feature that is crucial for its insulating capabilities and spongy texture. The pronounced crimp observed in finer fibres arises from a clear distinction between ortho-cortical and para-cortical cells, whereas in coarser fibres, this distribution is more irregular, resulting in reduced crimp. A comprehensive examination of the cortex includes the following structures.

    a) **Cell membrane complex:** The cell membrane complex (CMC) envelops and binds the cortical cells together, facilitating dye absorption and contributing to the structural integrity of the wool. Composed of proteins and waxy lipids, it functions similarly to mortar, preserving the fibre's integrity and enabling dye uptake. Nevertheless, the weak bonds within the CMC render it susceptible to damage from abrasion and harsh chemicals, which can compromise the wool over time.
    b) **Macrofibril:** Within each cortical cell, macrofibrils are long, filamentous structures. These macrofibrils consist of bundles of even finer filaments known as microfibrils.
    c) **Matrix:** The matrix surrounds the microfibrils and is rich in proteins that are high in sulphur content. These proteins have a strong affinity for water, enhancing the absorbency of wool, allowing it to retain up to 30% of its weight in water and absorb substantial amounts of dye. Additionally, the matrix contributes to the fire resistance and anti-static properties of wool.
    d) **Microfibril:** Situated within the matrix, microfibrils impart strength and flexibility to the wool fibre, akin to the role of steel rods in concrete.
    e) **Twisted molecular chains and helical coils:** Microfibrils are composed of twisted molecular chains that exhibit helical coiling,

which is further strengthened by hydrogen and disulfide bonds. This configuration endows wool with flexibility, elasticity, and resilience, enabling it to maintain its shape and resist wrinkling.

3. **Medulla:** The central core of the fiber extends lengthwise and constitutes 10-80% of the fiber's volume.

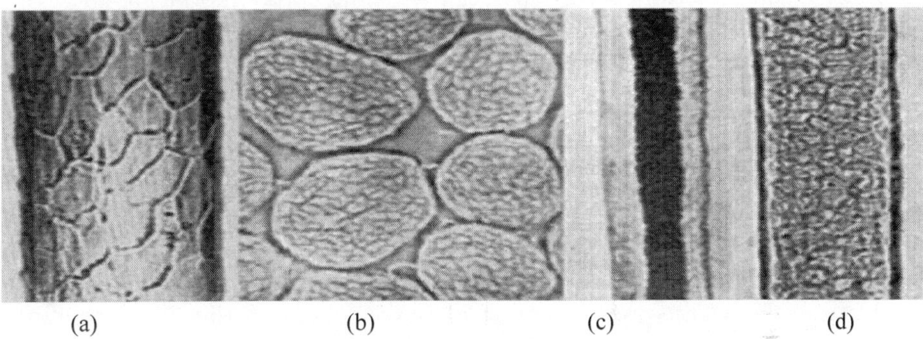

Figure 38: Structure of wool

a) Overlapping scales of cuticle or epidermis cells giving serrated surface appearance
b) Fiber cross sections showing cortex
c) Longitudinal cross section showing hollow medulla
d) Longitudinal cross section showing medulla lattice

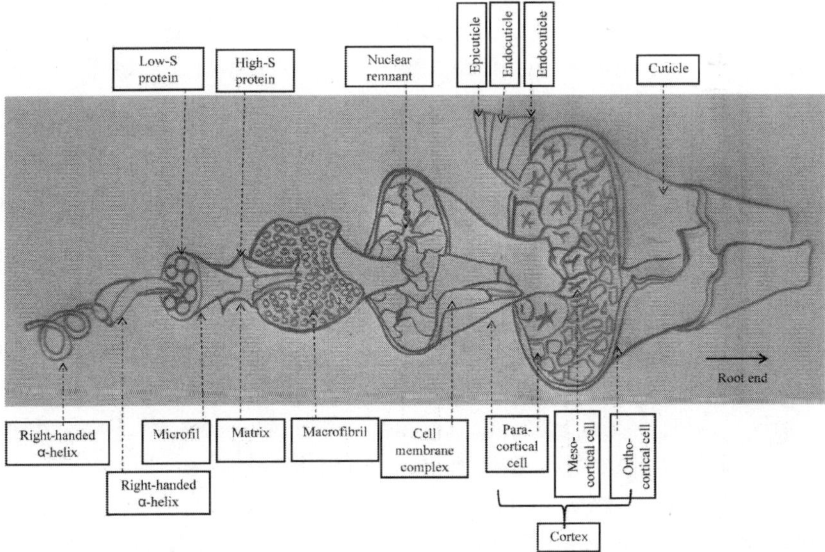

Figure 39: Detailed structure of wool fibre

Chemical Structure of Wool: The unique properties of wool arise from its chemical composition, which influences its texture, elasticity, staple length, and crimp. Wool is primarily composed of protein polymers created from amino acids, along with trace amounts of fat, calcium, and sodium. Additionally, raw wool contains wool grease and suint, a water-soluble substance derived from perspiration. Wool comprises over 170 distinct proteins that are unevenly distributed throughout the fibre, resulting in diverse physical and chemical characteristics in various regions. The proteins found in wool are constructed from amino acids, which are defined by their basic amino (-NH) and acidic carboxyl (-COOH) functional groups. Wool includes 18 of the 22 naturally occurring amino acids. These amino acids vary in their side chains, which may be hydrophobic (repelling water), hydrophilic (attracting water), acidic, basic, or contain sulphur. In wool, amino acids are linked to create extensive polymer chains known as polypeptides. These polypeptides are classified as polyamides due to the presence of amide linkages connecting their structural units. The repeating unit of amide (-NHCHRCO) is referred to as a peptide.

Within the wool structure, polypeptide chains are interconnected by various covalent bonds (crosslinks) and non-covalent interactions, which contribute to the overall integrity of the protein. The most prominent crosslinks in wool are the disulfide bonds containing sulfur, which are established during the growth of the fibre through a process known as keratinization. These disulfide bonds make keratin fibres insoluble in water and increase their resistance to both chemical and physical degradation compared to other proteins. Furthermore, they play a vital role in the "setting" phase of wool fabric finishing, where the molecular bonds are reorganized to provide smooth-drying characteristics, thus reducing the necessity for ironing post-wash.

Another form of crosslink is the isopeptide bond, which occurs between amino acids that possess acidic or basic groups. Beyond these chemical crosslinks, the stability of wool is further augmented by non-covalent interactions among the side groups of amino acids. Hydrophobic interactions take place between hydrocarbon side groups, while ionic interactions, also known as "salt linkages," occur between acidic (carboxyl) and basic (amino) side chains. These ionic interactions are essential for maintaining the stability of the fibre in both wet and dry environments. Additionally, wool exhibits amphoteric characteristics due to the presence of carboxyl and amino groups, enabling it to absorb and release both acids and bases, which aids in pH buffering.

Physical Structure of Wool: Wool possesses a complex physical structure, categorizing it as a biological composite with unique chemical and physical regions. Australian merino wool fibres, typically ranging from 17 to 25 micrometres in diameter, are composed of two primary cell types: internal cortex cells and external cuticle cells. The cortex, which constitutes approximately 90% of the fibre, is made up of overlapping spindle-shaped cells. These cells are interconnected by the cell membrane complex (CMC), a layer of lightly cross-linked proteins and waxy lipids that serves to separate the cortex from the cuticle. Although the CMC accounts for only about 5% of the fibre's mass, it is integral to the properties of wool. It is mechanically relatively weak, which means that during prolonged use, wool fabrics frequently deteriorate along the interfaces between cortical cells, resulting in fibrillation.

Fine wool fibres, such as those derived from merino sheep, exhibit two types of cortical cells (ortho and para) arranged bilaterally. This configuration contributes to the desirable natural crimp found in merino wool fibres. In coarser wool (exceeding 25 micrometres in diameter), the differentiation between these cell types is less distinct.

The protein composition in wool differs across various fibre regions. Certain proteins within the microfibrils exhibit a helical structure, which imparts flexibility, elasticity, resilience, and excellent wrinkle recovery to wool. Conversely, the proteins found in the matrix that encases the microfibrils possess a more amorphous configuration. This matrix enables wool to absorb as much as 30% of its weight in moisture without feeling damp, thereby enhancing wool's versatility, which ranges from billiard cloths to fine textiles. Furthermore, the matrix facilitates the retention of a significant quantity of dye, thereby augmenting its colour and functionality.

Comparison of Wool and Hair: Wool, a fibre that has evolved over millennia to provide protection and insulation for sheep, stands as the most complex and versatile of all textile fibres. Its absorbent characteristics render fine wool products remarkably comfortable to wear. Moreover, the chemical structure of wool permits it to be dyed easily in a broad spectrum of hues, from gentle pastels to rich, vibrant shades. This versatility truly earns wool the designation of "Nature's Wonder Fiber."

A wool fibre comprises two distinct cellular layers. The outer protective sheath features flat, irregularly shaped scales akin to those of fish, which overlap one another. These scales are more loosely attached to the inner layer compared to hair. The outer layer of scales plays a crucial role in determining the quality of wool due to its effects on shrinking, strengthening, and felting conditions. Under natural conditions, the outer wool scales are coated with wool sweat (grease) or yolk, known as suint, which is secreted by specialized glands to maintain the fibre's condition. In fact, the term 'condition' refers to the amount of grease or oil present in the wool. A portion of suint is water-soluble and can be eliminated during washing. The inner layer of the wool fibre, known as the cortex, consists of long fibrils that are cemented together. In contrast, hair encloses an air-filled medulla, resulting in hair possessing less strength than wool. The following table illustrates the comparison between wool and hair.

Table 22: Comparison of wool and hair

Sr. No.	Parameters	Wool	Hair
1	Medulla	Almost absent	Present and pronounced
2	Cuticle	Irregular	Regular and smooth
3	Sides	Scaly projections	Smooth
4	Diameter	Less	More
5	Growth	Continuous, if not sheared	Reaches a maximum and then shed
6	Softness	More	Less
7	Elasticity	More	Less
8	Heat retention	More	Less
9	Moisture retention	More (12-17%)	Less (7-12%)
10	Dye retention	Permanent	Temporary
11	Lustre	More	Less
12	Inflammability	Less	More

14

Properties of Wool

Wool, which is sourced from sheep, is a significant protein fibre categorized as an animal fibre. Its hygroscopic nature allows it to absorb moisture effectively. Composed mainly of keratin, a protein created by amino acids linked through peptide bonds, wool possesses unique characteristics due to its sulphur content, setting it apart from other animal fibres. The fibres of wool are crimped and elastic, growing in clusters. This crimping generates air pockets that provide insulation, while the outer surface features overlapping, serrated scales that enable the fibres to interlock and create felt. For thousands of years, wool has been utilized as a textile because of these distinctive attributes.

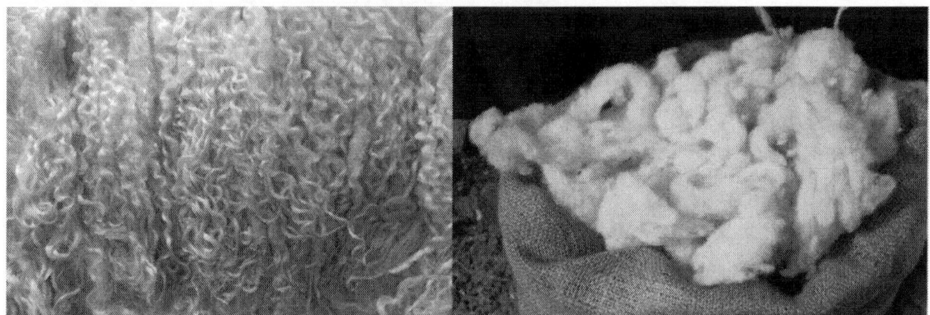

Figure 40: Wool fibre

Properties of Wool Fiber: The properties of wool fibre include the following:

1. **Composition of amino acids:** Wool fibres consist of amino acids that form the protein keratin.
2. **Absorbency:** Wool exhibits high absorbency, which allows it to effectively take in moisture.
3. **Moisture retention:** Wool can hold a considerable amount of moisture in relation to its weight.
4. **Thermal insulation:** Wool fibres offer superior warmth compared to many other fibres due to their crimp and structure.
5. **Chemical resistance:** Wool fibres show limited resistance to alkalis (such as strong soaps or detergents) but are resistant to acids.

6. **Elasticity and resilience:** Wool is recognized for its excellent elasticity and its ability to revert to its original shape after being stretched.

To process wool fibres effectively, it is essential to comprehend both their physical and chemical properties. This understanding aids in identifying the optimal methods for processing and treating wool to preserve its quality and functionality.

A. **Physical properties of wool fibre:** The following characteristics render wool a versatile and valuable fibre for various applications.

1. **Specific gravity:** The specific gravity of wool ranges from 1.31 to 1.32.
2. **Density:** The density of wool ranges from 1.31 to 1.032 g/cc.
3. **Thickness:** The thickness of wool ranges from 16 to 40 micrometres.
4. **Length:** The length of wool ranges from 35 to 250 mm.
5. **Colour:** The hue of wool can be white, nearly white, brown, or black.
6. **Flame reaction:** Upon exposure to flame, wool emits a characteristic scent reminiscent of burnt horn or hair. This phenomenon occurs because wool is a protein-based fibre, and its combustion releases sulphur compounds along with other byproducts linked to the keratin present in the fibres.
7. **Lustre:** Coarse wool fibres generally exhibit a greater lustre than their finer counterparts. This is attributed to the larger diameter of coarse fibres, which reflects more light, resulting in a shinier look.
8. **Moisture regains:** Wool fibres possess a moisture regain of 13-16%, signifying their considerable absorbency. They can retain a significant amount of moisture in relation to their weight. Nevertheless, the strength of wool diminishes when wet, potentially impacting its durability during handling or laundering. Despite this drawback, wool maintains its insulating characteristics even when moist. Furthermore, wool is prone to shrinkage when washed, particularly if subjected to hot water or agitation. Careful maintenance is essential to preserve its dimensions and form.
9. **Electrostatic reaction:** When dry, wool fibres are prone to becoming highly electrostatic, which can result in static cling and the attraction of dust and particles. This occurs due to friction and the accumulation of electrical charges on the fibres' surface.
10. **Strength:** Among all natural fibres, wool is the weakest. The strength of wool fibre can be enhanced through the use of ply yarns. A tightly twisted two-ply yarn is often seen as a guarantee of durability. Similarly, tightly twisted single yarns contribute to the creation of a robust fabric.

- **Tenacity:** The tensile strength of wool in its dry state ranges from 1-1.7 g/d (grams per denier), while in its wet state, it ranges from 0.8 to 1.6 g/d (weaker when wet due to a reduction in hydrogen bonds).
- **Tenacity range:** 3 to 30 cN/tex (centinewtons/tex).

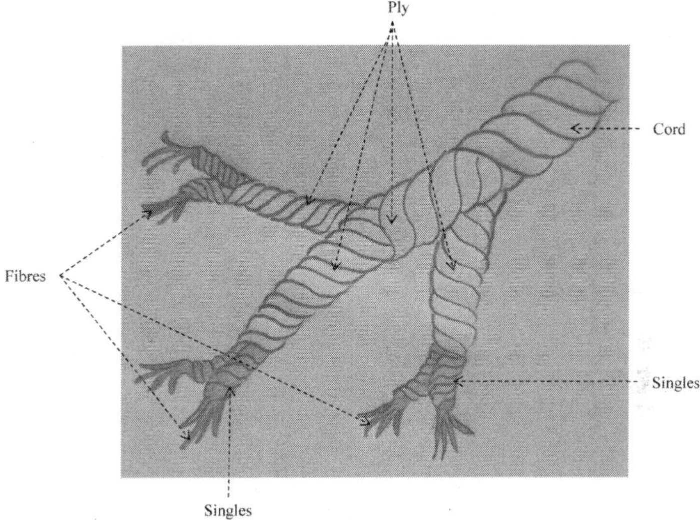

Figure 41: Ply yarns

11. **Elasticity:** The extent to which wool can be stretched varies based on its quality, with fibers capable of elongating by 25-30 percent of their original length. This characteristic mitigates the risk of tearing when under tension and facilitates unrestricted body movements. Additionally, chemical treatments enhance the retention of shape.
    - **Breaking extension:** 42.5%
    - **Recovery:** 69% at 5% extension.
12. **Elongation at break:** This property is crucial for wool, indicating the extent to which it can stretch before fracturing, expressed as a percentage of its initial length.
    - **Standard (Dry) Condition:** Under typical dry conditions, wool exhibits a standard elongation at break, with fibers being relatively elastic but prone to breaking after stretching approximately 25-35% of their original length.
    - **Wet Condition:** In a wet state, wool fibers can stretch further due to moisture absorption, with elongation at break increasing to allow stretching of 25-50% of their original length before breaking.

13. **Feel on skin:** Wool is generally perceived as soft to the touch, enhancing its comfort and wearability. The degree of softness may vary based on the fineness and processing of the wool, yet it consistently offers a pleasant texture against the skin.
14. **Resiliency:** Wool fibers exhibit a high level of resilience. Quality wool is characterized by its softness and resilience, attributed to its crimp structure. In contrast, lower quality wool may feel harsh. Compared to many other fibers, such as cotton or linen, wool fibers are less prone to wrinkling, a result of their inherent elasticity and crimp.
15. **Abrasion resistance:** Wool fibers possess commendable abrasion resistance, enabling them to endure wear and friction more effectively than numerous other fibers. This durability renders wool suitable for diverse applications, including garments and carpets.
16. **Dimensional stability:** Wool exhibits limited dimensional stability, which implies that it may alter in shape and size over time. It is susceptible to felting, particularly when subjected to heat, moisture, or agitation. Felting transpires when the scales on the wool fibers interlock, resulting in the fabric shrinking and becoming denser. This phenomenon can induce alterations in the garment's dimensions and texture. Adequate care, such as gentle washing and avoiding elevated temperatures, can assist in alleviating these concerns.
17. **Effect of heat:** Heat has a considerable impact on the fiber. Wool becomes coarse at 100°C and starts to decompose at slightly elevated temperatures. It possesses a plastic quality that enables it to retain shape at melting temperatures.
18. **Effect of sunlight:** Exposure to sunlight can lead to the discoloration of wool and a gradual increase in its harshness. Ultraviolet rays degrade the fibers, resulting in fading and a coarser texture. Proper storage, away from direct sunlight, can aid in maintaining the appearance and softness of wool.

B. Chemical properties of wool fiber:

1. **Effect of acids:** Wool is vulnerable to hot concentrated sulfuric acid, which causes complete decomposition. Generally, it exhibits resistance to mineral acids of all strengths, even at elevated temperatures, although nitric acids can inflict damage through oxidation.
2. **Effects of alkalis:** The chemical composition of wool keratin renders it particularly sensitive to alkaline substances. Wool will dissolve in caustic soda solutions that have minimal effects on cotton. Strong alkalis

exert a significant impact on wool fibers, while weak alkalis do not affect wool.

3. **Effect of resistance to compression:** Resistance to compression (R to C) quantifies the force required to compress wool to a designated volume. This characteristic is crucial in assessing wool's appropriateness for various applications, as it correlates with the fiber's diameter and crimp properties. Wool exhibiting higher resistance to compression can retain its shape and offer superior cushioning, making it suitable for applications such as upholstery or carpeting.
4. **Impact of resilience:** Wool fibers demonstrate significant resilience, allowing them to revert to their original shape after being stretched. When wet, these fibers can extend up to 50% of their initial length, while in a dry state, they can stretch up to 35%. This remarkable elasticity aids wool garments in maintaining their shape and recovering from stretching, thereby enhancing their durability and comfort.
5. **Impact of organic solvents:** Wool generally exhibits resistance to most organic solvents. In contrast to certain fibers that may deteriorate or alter when exposed to solvents such as acetone or alcohol, wool usually preserves its structure and characteristics. Nonetheless, it is advisable to test solvents on a small, inconspicuous area beforehand to confirm that there are no negative effects.
6. **Impact of insects:** Wool is vulnerable to damage from specific insects, particularly the larvae of moths and beetles that consume wool. These pests can inflict considerable harm by eating the protein fibers, resulting in holes and degradation of woollen fabrics. To safeguard wool items, it is crucial to implement preventive strategies such as appropriate storage, the use of moth repellents, and regular cleaning.
7. **Impact of bleach:** Chlorine bleach poses a threat to wool, as it can weaken the fibers, lead to discoloration, and compromise the fabric's structure. Instead, wool is typically bleached with alternative agents like potassium permanganate ($KMnO_4$) or sodium peroxide (Na_2O_2), which are more effective in preserving the integrity of the wool fibers while achieving the desired lightening effect.
8. **Impact of microorganisms:** Wool is prone to mildew if it remains damp or wet for prolonged periods. Microorganisms, especially mould and mildew, can proliferate on wool fibers under such conditions, resulting in unpleasant odours, discoloration, and deterioration of the fabric. To avert mildew, it is vital to keep wool garments and textiles dry and well-ventilated.

9. **Dyeing capability:** Wool possesses remarkable dyeing capability, allowing it to absorb a diverse array of dyes thoroughly and uniformly. This property enables wool to attain vibrant, rich hues without requiring additional chemicals or mordants. Its strong affinity for dyes renders wool a favoured option for creating beautifully coloured textiles.
10. **Impact of colourfastness:** The colourfastness of wool is typically commendable, demonstrating a strong capacity to retain dyes effectively. Acid dyes, chrome dyes, and mordant dyes are frequently employed for wool, as they bond efficiently with the fiber. Dye molecules are drawn to the amorphous regions within the wool structure, guaranteeing deep and enduring colours.

15

Classification Sorting and Grading of Wool

Classifying wool is crucial in the textile sector for upholding quality standards, enhancing processing techniques, and ensuring that wool satisfies particular market demands. Precise classification aids in choosing the appropriate wool type for a variety of products, ranging from apparel to carpets, and guarantees uniformity and performance across diverse applications.

Classification of Wool: Wool is categorized in a manner akin to hides and skins, based on its intended application. It is divided into three primary types according to fibre length: combing wool, clothing wool, and carpet wool. Long fibres are combed and twisted to produce worsted yarn, whereas shorter fibres are carded and blended in various directions to create woollens. Carpet wool is characterized by an average fineness of 30 to 50 μm and a medullation percentage between 15% and 68%. The quality of wool is influenced by factors such as breeding conditions, climatic conditions, diet, and overall care. For example, excessive moisture can strip natural grease, while colder climates lead to tougher, heavier fibres. Consequently, wool can be classified in two distinct manners:

A. By the sheep from which it is sourced, and

B. By fleece

A. Classification by sheep: Wool is categorized based on the sheep from which it is sheared, as outlined below.

1. **Merino wool:** Merino sheep, which originated in Spain, yield some of the highest quality wool. This wool is robust, fine, and elastic, with a relatively short length of 1 to 5 inches (25 to 125 mm). Merino wool is recognized for its high crimp and numerous scales, which enhance its warmth and spinning characteristics. Owing to these attributes, Merino wool is highly sought after for premium wool garments.

2. **Class-two wool:** This category of sheep, which hails from England, Scotland, Ireland, and Wales, produces fibres that are comparatively strong, fine, and elastic, with lengths ranging from 2 to 8 inches (50

to 200 mm). These fibres exhibit a high density of scales per inch and possess good crimp, contributing to their durability and quality.

3. **Class-three wool:** This category of sheep, which hails from the United Kingdom, yields coarser wool fibers that measure approximately 4 to 18 inches in length. In comparison to earlier types, these fibers exhibit fewer scales and reduced crimp, yet they possess a smoother and more lustrous quality. Although they are less elastic and resilient, this wool remains of good quality and is utilized in the production of clothing.
4. **Class-four wool:** Commonly known as mongrel or half-breed sheep, this class produces wool fibers that vary in length from 1 to 16 inches (25 to 400 mm). These fibers are coarse and resemble hair, featuring relatively few scales and minimal crimp. They are generally smoother and more lustrous but lack elasticity and strength. Due to these properties, this wool is considered less desirable and is mainly employed for carpets, rugs, and lower-grade clothing.

B. **Classification by fleece:** The process of shearing involves the removal of a sheep's wool fleece, typically conducted in the spring, although the timing may differ by region. Sheep are not washed prior to shearing, although they might be treated with an antiseptic bath if mandated by regulations. The classification of wool based on fleece is as follows:

1. **Lamb's wool:** The fleece obtained from a lamb aged six to eight months during its initial shearing is referred to as lamb's wool. Also known as fleece wool or first clip, this wool is characterized by its natural, tapered ends, which enhance its softness.
2. **Hogget wool:** Harvested from sheep that are approximately twelve to fourteen months old and have not been shorn previously, hogget wool is fine, soft, resilient, and features tapered ends. It is regarded as mature wool and is primarily utilized for the production of warp yarns.
3. **Wether wool:** Wether wool is sourced from sheep older than fourteen months and is sheared after their first fleece has been removed. These fleeces tend to be more soiled and dirtier, having accumulated more debris since the initial shearing.
4. **Pulled wool:** Pulled wool is derived from the pelts of sheep that have been slaughtered for meat. The extraction of the wool involves the use of chemicals, which compromises the quality of the fibers. Consequently, pulled wool yields a lower-grade fabric.

5. **Dead wool:** This category of wool, known as dead wool, is sourced from sheep that have died due to old age or have been accidentally killed. It is important to distinguish this from pulled wool. Dead wool is of substandard quality and is generally utilized in the production of low-grade textiles.
6. **Cotty wool:** Cotty wool is harvested from sheep that have been subjected to extreme weather conditions, adversely affecting the quality of the fleece. This type of wool is often characterized by its poor grade, hardness, and brittleness as a result of the severe environmental influences.
7. **Tag locks:** Tag locks denote the torn, ragged, or discoloured sections of a fleece. These portions are usually sold separately and are regarded as an inferior grade of wool.

Sorting of Wool: Wool constitutes less than 5% of the global fiber market and is costly to produce and process, necessitating alignment with consumer demand to maintain economic viability. Sheep producers engaged with wool breeds must be knowledgeable about wool technology, mill specifications, and pricing structures. Understanding wool grades is crucial for addressing significant concerns such as genetic consistency and efficient wool harvesting and preparation. Wool is categorized into utility classes based on fiber fineness and diameter, with finer fibers resulting in lighter materials and thicker fibers leading to heavier products. In the sorting process, wool is segmented according to quality from various body regions to minimize variation and enhance market value. The highest quality wool is designated for clothing, while lower quality or second fleece is allocated for rugs. Grades are determined by type, length, fineness, elasticity, and strength.

Upon thorough analysis, the fleece of a typical sheep exhibits thirteen or fourteen distinct qualities of wool. Generally, each quality is located in the same region of every fleece from that particular sheep species. To maintain consistent wool quality, we categorize each fleece into sections that correspond to these various qualities. Given that each fleece yields only a limited quantity of each quality, numerous fleeces are required to create a uniform type of yarn or fabric. In factories that handle thousands of fleeces weekly, this sorting procedure guarantees a consistent fiber quality for both premium and lower-grade products.

Woolen and worsted sorting terminology: Each thread has its own method of fleece sorting and specific terminology associated with it.

- In woolen sorting, the emphasis is placed on fiber fineness, leading to the following classifications:

- **Picklock (Fore shoulder):** The finest in fiber quality, elasticity, and staple strength.
- **Prime (Middle of body):** Slightly less strong, yet comparable in other aspects.
- **Choice (Back):** Accurate, but not as fine as prime.
- **Super (Lion):** Less valuable than choice, yet similar in general characteristics.
- **Head:** Inferior wool types sourced from this area of the sheep.
- **Drawn right (Lower sides):** Exhibiting softness, yet fair wool quality.
- **Seconds (Throat and breast):** The best wool from these regions.
- **Breech:** Short, coarse hair from the rear sections.

- The worsted selection prioritizes fineness, as indicated by the following classification:
  - **Blue:** Sourced from the neck.
  - **Fine:** Taken from the shoulders.
  - **Neat:** Derived from the middle of the sides and back.
  - **Brown-drawing:** Collected from the haunches.
  - **Breech or Britch:** Obtained from the tail and hind legs.
  - **Cow-tail:** Very robust wool from the rear legs.
  - **Brokes:** Sourced from the belly and lower front legs, categorized as super, middle, and common based on quality.

English Wool Sorting: For manufacturers aiming to grade wool finely, the fleece of the common sheep is mapped with the lowest numbers representing the highest qualities.

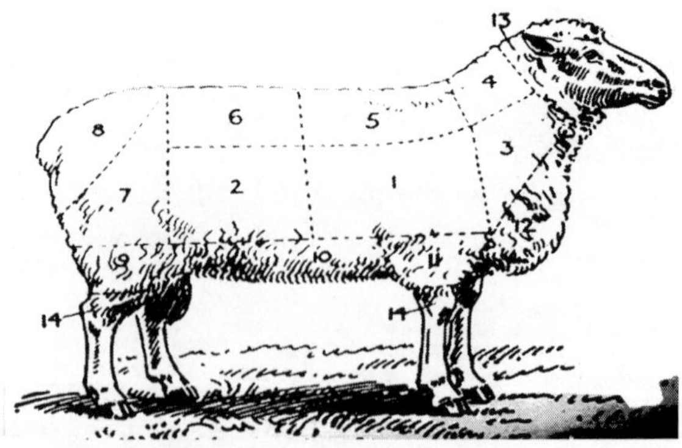

Figure 42: English wool sorting

1. **Shoulder:** Wool that is long and fine, growing close and uniformly.
2. **Side:** Robust wool, comparable in quality to that of the shoulder.
3. **Neck:** Wool that is short yet fine, occasionally blended with greyish fibers near the head.
4. **Back of neck:** Of lesser quality compared to the shoulder, side, and neck.
5. **Top of fore shoulder:** Defective and irregular, though of moderate quality.
6. **Lions and back:** Wool that is coarse and short, yet relatively consistent in nature.
7. **Middle of haunch:** Long, robust wool exhibiting a range of qualities from fine to coarse.
8. **Hinder parts:** Wool that is coarse and long, which may exhibit hairiness.
9. **Top of hind legs:** Comparable to the middle of the haunch, albeit dirtier.
10. **Under body:** Wool that is short and dirty, finer near the fore legs; referred to as 'brokes.'
11. **Top of fore legs:** Wool that is both short and fine.
12. **Throat:** Wool that is irregular, short, and kempy, frequently containing grass and fodder.
13. **Head:** Wool that is short, rough, and coarse.
14. **Shanks:** Wool that is rough, hard, and very short, possessing minimal value.

Sorting the Merino Fleece: The Merino sheep produces the finest wool, and its fleece is carefully graded as follows:

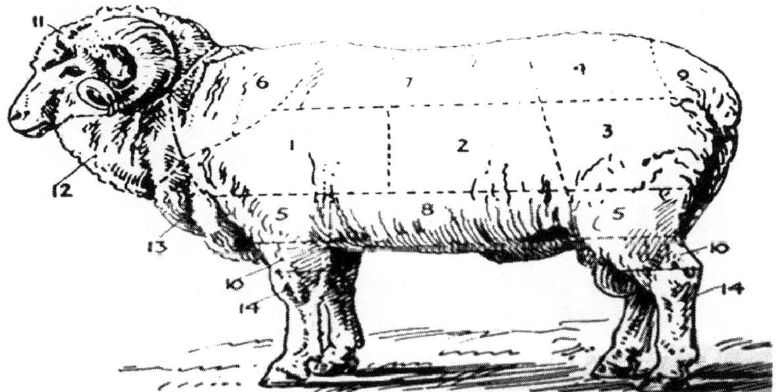

Figure 43: Merino wool sorting

1. **Shoulders:** The finest wool, recognized for its durability, long staple length, soft feel, and consistent quality.
2. **Sides:** Comparable to the shoulders in terms of quality, robust and soft with a consistent texture.
3. **Lower part of back:** Wool of good quality with a staple length akin to that of the shoulders and sides, though not as soft or fine.
4. **Loin and back:** Characterized by a shorter staple and coarser hair, generally possessing true characteristics but occasionally tender.
5. **Upper parts of legs:** Wool that is moderately long yet coarse, often falling in loose locks and sometimes containing plant matter.
6. **Upper portion of neck:** Wool of inferior quality, irregular and faulty, frequently mixed with thorns, twigs, and grass.
7. **Central part of back:** Wool similar to that of the loin and back, but tends to be more tender.
8. **Belly:** Short, unclean, low-quality wool that is somewhat tender.
9. **Root of tail:** Coarse, short, and shiny wool, often mixed with kemps or dead hairs.
10. **Lower parts of legs:** Wool that is dirty, greasy, and rough, lacking in fineness and curliness, typically containing burrs and plant matter.
11. **Head:** Stiff, straight, coarse wool mixed with fodder and kempy fibers.
12. **Throat:** Similar to the head, characterized by stiffness and coarseness.
13. **Chest:** Comparable to the head, stiff and coarse in texture.
14. **Shins:** Short, straight, stiff wool with minimal value for textile production.

Spanish Shorting: The Merino fleece is categorized by the Spaniards into four sections:

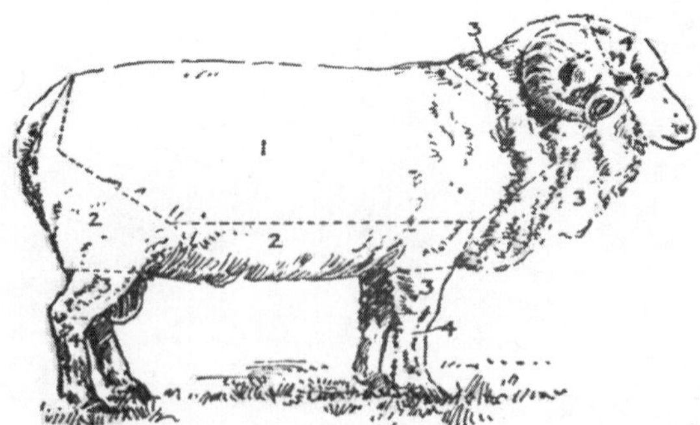

Figure 44: Spanish wool sorting

1. **Refina:** Encompasses wool from the lower jaw to the fore shoulder, extending across to the haunch and curving around to the back above the tail.
2. **Fina:** Wool sourced from the belly, hindquarters, and upper thighs.
3. **Tercina:** Short wool derived from the head, throat, lower neck, and shoulders, concluding at the joint.
4. **Inferior:** Wool obtained from the forehead, cheeks, tail, and legs.

Grading of Wool: The systems for grading wool assess its quality and appropriateness for various applications, including garments and carpets. Wool is evaluated based on several criteria such as type, length, fineness, elasticity, and strength. The three primary grading systems utilized are:

- American system or blood system
- English/British system/numerical system/spinning count system
- Micron system

A. **American system or blood system:** Established in the early 1800s, the American wool grading system is of U.S. origin and predominantly employed in the United States. This system is predicated on the fineness of Merino wool. The wool grade is determined by the percentage of Merino blood present in the sheep, which typically correlates with the fineness of the wool produced. The American wool classification comprises seven grades, with fiber diameter categorized as fine, ½-blood, 3/8-blood, 1/4-blood, low 1/4-blood, common, and braid.
   1. First quality wool is recognized as fine, corresponding to the quality obtainable from a full to three-quarter-blood Merino sheep.
   2. Second quality wool is akin to that derived from a half-blood Merino.
   3. The lowest quality grades are termed common and braid; these are characterized by coarseness, minimal crimp, a limited number of scales, and a somewhat hair-like appearance.

B. **English/British system/numerical system/spinning count system:** The grading framework prevalent in the global market is founded on the British numbering system, which connects the fineness or diameter of wool fibers to the type of combed or worsted yarn that can be produced from one pound of scoured wool. The English grading system offers narrower ranges and a more precise terminology compared to the American system. It employs a measurement known as the 'spinning count,' which is based on the number of 'hanks' of yarn that can be spun from one pound of clean wool using the machinery available at the time of the system's inception. As wool becomes finer, a greater number of hanks or yards of yarn can be produced from a pound of clean wool, resulting in a higher spinning count. Theoretically, one pound of clean wool with a spinning count of 62s could yield 62 hanks or 104,160 feet of yarn (with a hank measuring 560 yards in length). The English or Spinning Count system for grading wool assigns a numerical value to its fineness.

1. The highest quality wool is that which is sufficiently fine to be spun into the top yarn counts of 80s, 70s, and 64s (indicating the number of 560 yards in one pound).
2. The second quality is fine enough to be spun into yarn counts of 62s, 60s, and 58s.
3. The lowest grade can only be spun into yarn counts of 40s and 36s. The English or spinning count grades of wool that are commonly utilized today include: 80s, 70s, 64s, 62s, 60s, 58s, 56s, 54s, 50s, 48s, 46s, 44s, 40s, and 36s. Wool is graded based on the diameter of the yarn it can produce; for instance, clean wool weighing 0.454 kg (1 lb) of 80's can be spun into 80 hanks of wool yarn (equivalent to 25 miles or 40 km), while a 36's coarse wool can yield 36 hanks (11 miles or 17.6 km) per 0.454 kg (1 lb).

Wool classification is also influenced by length, which is categorized into three classes:

- **Combing:** This class encompasses wool lengths ranging from 2.5 inches to 7 inches, suitable for spinning and weaving into durable worsted fabrics.
- **French combing:** This wool is longer than that used for clothing but shorter than combing wool.
- **Clothing:** This wool is too short to be spun into thread but can be utilized to create woolen fabrics that do not retain press or last as long as worsted suits.

British wool grades include

- **Diamond wool:** The finest white British wool, measuring only 1 to 1.5 inches in length.
- **Extrapick/pick:** This wool is less fine than diamond wool, with a length of about 1 inch.
- **Shafty:** This refers to pick teg wool with additional length.
- **Super:** This is a less fine wool measuring 4 to 5 inches in length.

- **Half-bred:** Wool that is coarser than super, measuring 6 to 7 inches long.
- **Deep half-bred:** This wool is longer and coarser.
- **Lustre:** This wool is longer, curly, and possesses a natural sheen.
- **Roller:** Lustre Hogg wool, measuring 12 to 15 inches long, is regarded as a specialty grade.
- **Cotts:** This refers to matted wool.
- **Arable:** Wool containing dusty soil.
- **Cash:** Fleece with bad faults.

Table 23: Comparative wool grading table of United States system and British system

United States System	British System
Fine (full-to-three-quarter-blood)	80s, 70s, 64s
Half - blood	62s, 60s, 58s
Three - eights - blood	56s
Quarter - blood	50s, 48s
Low - quarter - blood	46s
Common	44s
Braid	40s, 36s

The Micron System: The most technical and precise grading system has been developed by the Denver Wool Laboratory, USDA. This system places a greater emphasis on a detailed and exact method for describing wool grades, resulting in a measuring system that accurately assesses individual fibers. Fineness is represented as the average fiber diameter. The system categorizes wool into 16 distinct categories based on the average fiber diameter measured in micrometers. The unit of measurement is the micron, equivalent to one millionth of a meter or 1/25,000 of an inch. Fineness is articulated as the mean fiber diameter. Ultimately, this system is expected to become the standard for wool description in the United States.

Grading of Indian Wools: The primary wool grading centres in India are located in Jaipur, Bikaner, and Jodhpur. The grading system employs single letters A, B, C, and D to denote fineness, which corresponds to various grades. The grading process considers factors such as length, fineness, colour, and the presence of vegetable matter or burr. The fundamental criteria of the Indian wool grading system include:

- A visual grading system that assesses fiber length and diameter.
- Differentiation of wool based on fineness.

- Grading conducted at the point of shearing.
- Wool is graded separately after skirting into coarse, fine, and medium categories.

Length of wool fiber: A longer fiber length correlates with a higher grade. Fiber length is measured in hanks, with staple length referring to the length of an unstretched lock of shorn wool. A greater staple length indicates a superior grade. Double letters signify wool with a long staple length, while single letters A, B, C, or D indicate that the staple length exceeds 2 ½ inches.

Fineness: The fineness of wool is determined by the diameter of the wool fibers. The thickness of the wool fiber is measured in microns (one millionth of a meter = μ). Based on fineness, wool can be classified as fine, crossbred, medium, or long coarse. The grading system utilizes a numerical range from the 30s to the 80s, which indicates the relative diameter of the wool fibers (fineness). The greater the number, the finer the wool. The quality number signifies the potential theoretical spinning count within the worsted system. The quality index number reflects the range of fibre diameters to prevent comparisons of wool quality with the associated quality number.

Table 24: Grading of wool on the basis of fineness of wool fibre

Grade	Fineness (spinning count)	Fineness (Fibre diameter)
Super A- Super-fine quality	> 58's	< 25 microns
A - Fine wool quality	54-56's	< 34.4 microns
B - Medium wool quality	46-50's	34.4-37.4 microns
C - Strong wool quality	40-44's	37.4-40.4 microns
D - Coarse wool quality	< 40's	> 40.1 microns

Colour: White colour of wool indicated by assigning the letters ALB. Whereas yellow colour is indicated as LY (Light Yellow), MY (Medium Yellow) and HY (Heavy Yellow). Black or brown colour has no place under grading.

Vegetable content: On the basis of burr content, it may be of three types.

- LB for light burr content; below 3%
- MB for medium burr content; between 3-6%
- HB for heavy burr content; above 6%

Wool having 4 categories of length, 4 categories of fine-ness, 4 categories of colour and 3 categories burr. Thus, Indian wool can be classified into 4X4X4X3= 192 grades of wool.

Table 25: Wool grading system of Indian Standard Institute (I.S.I.)

Category	Fineness	Length	Vegetable content	Colour
Category A	< 34.4 microns	Above 75 mm	LB < 3.0%	White
Category B	34.4-37.0 microns	Below 75 mm	MB 3.0-6.0 %	Tinged white (TW)
Category C	37.1-40 microns	Below 75 mm	HB >6.0%	Light yellow (LY)
Category D	> 40.1 microns	Below 75 mm	HB >6.0%	Heavy yellow (HY)

Wool grading system (F.A.O.): Followed in states like Punjab, Haryana, Jammu and Kashmir, Gujrat and Karnataka.

Table 26: Wool grading system (FAO)

Code No.	Style description
Good	Wool with good colour, washed, skirted, light burr (0-3%) with reasonable quality and strength.
Good - average	Wool having yellow colour, well skirted, light burr (0-3%) with reasonable quality and strength.
Average	Wool of average washing, well skirted, medium burr (3-6%).
Inferior	Wool of poor washed category, heavy burr content >6.0%.

16

Quality and Impurities of Wool

Wool is a natural fiber that possesses numerous outstanding characteristics, rendering it both functional and visually appealing. Nevertheless, not all wool is created equal; there exists considerable variation in the quality found in the marketplace. This explains the significant fluctuations in price, as some wool is of superior quality and therefore commands a higher price. Various factors influence the quality of this fiber and, consequently, its value. Below is a list of some of these factors.

a) The diameter of the wool fiber is the most critical factor affecting its quality and price. Generally, merino wool is regarded as the highest quality due to its fine texture, which produces an exceptionally soft fiber. In simple terms, the finer the wool, the greater its worth.

b) Consistency in fiber diameter is another vital aspect; the more uniform the fiber, the higher its value.

c) The colour of wool also plays a significant role in determining its value. Wool that is white or off-white can accommodate a wider array of dyes used in fabric manufacturing.

d) Staple strength is a measure of the fiber's capacity to endure the production process. A higher staple strength results in less waste during manufacturing.

Wool is a valuable by-product of sheep, differentiated from hair by its enhanced elasticity, flexibility, and curliness. The quality of wool is affected by factors such as shrinking, strengthening, and felting conditions. In their natural form, wool fibers are coated with wool grease (sweat) and wool soap or yolk (suint), which are secreted by specialized glands to preserve the wool's condition. Wool is classified into three primary types of fibers:

1. **Fine wool fibers:** These fibers, such as those derived from Merino sheep, typically lack a medulla or hollow core and continue to grow throughout the sheep's lifespan. They are renowned for their softness and fineness.

2. **Hairs:** These are longer fibers that grow continuously and contain a medulla in certain sections of their length.

3. **Kemps:** These are short, coarse fibers that cease to grow at intervals and are ultimately shed from the fleece. They possess a medulla along their entire length and are remnants of the original outer coat.

When fleece is sheared from a sheep, it is categorized into various types based on its quality and characteristics. These classifications include:

1. **Merino or fine wool:** Renowned for its fine texture.
2. **Crossbred or medium wool:** A mixture of different breeds, providing medium quality fibers.
3. **Lustre long wool:** Noted for its sheen and longer fibers.
4. **Carpet wool:** Coarser wool generally utilized in carpet manufacturing.

Wool can also be sourced from the pelts of slaughtered animals, referred to as pulled wool, in contrast to shorn wool. Raw wool, or grease wool, contains natural grease, impurities from minerals, vegetable matter, and suint. The distinction between raw (grease) wool and clean wool is known as "shrinkage," which varies according to factors such as breed, diet, and soil. Buyers usually compensate for clean wool, factoring in the shrinkage in their assessments.

Quality Characteristics of Wool: The quality of wool is assessed based on several critical characteristics, which together account for roughly 80% of its value. These characteristics encompass colour, fiber length, fiber diameter, clean dry yield, and additional properties.

1. **Colour:** The colour of wool is significant because the presence of any pigment significantly diminishes the potential for dyeing. The optimal colour is white or creamy white with a noticeable lustre. Any pigment in the wool lessens its dyeing capabilities. Canary staining or yellowing of wool is a common issue in Indian wool, caused by bacterial activity in high pH and low grease conditions.
2. **Fiber diameter:** Fine wool has a fiber diameter of 17-18mμ (micron), while coarse wool may reach nearly double that (35 – 38mμ). Wools with a fiber diameter of less than 30mμ (non-medullated) are only appropriate for clothing textiles.
3. **Fiber length:** The length of fibers is a crucial factor from a spinning perspective. Longer fibers are favoured as they contribute to the production of stronger yarns. The length of fibers is affected by environmental conditions and the age of the sheep, with staple length typically evaluated after the sheep reaches 12 months of age. Fine wool fibers are generally shorter, often measuring less than 6 cm, yet this does not render them coarse. On the contrary, they are utilized for high-quality clothing textiles due to their softness and fineness. Coarse wool,

on the other hand, is characterized by a thicker diameter and typically longer fibers, which are more suitable for products like carpets rather than delicate garments.

4. **Clean dry yield:** Clean dry yield denotes the amount of pure wool obtained after the removal of natural grease, plant materials (such as grass seeds and burrs), and soil. These contaminants can constitute as much as 20 percent of the total weight.
5. **Strength:** The strength of the fiber is another significant attribute. Factors such as nutrition and health, particularly conditions like helminthiasis, can influence fiber strength. This property is evaluated by applying tension to the wool staple; fine wool is capable of enduring a sudden pull.
6. **Crimp:** The inherent waviness of wool fibers is referred to as crimp. The number of crimps per unit length affects the elasticity and texture of the wool.
7. **Elasticity:** Fine wool fibers can extend up to 70% beyond their original length before they break.
8. **Medullation:** Medullated fibers can be identified by sorting a sample against a black background, such as black velvet. In a laboratory setting, medullation is assessed by immersing a wool sample in xylol and examining it under high magnification.

These characteristics collectively assist in determining the suitability of wool for various uses, ranging from high-quality textiles to carpets. A comprehensive understanding and measurement of these traits are vital for both wool producers and buyers in evaluating the value and quality of wool.

Table 27: Characteristic of main types of wool

Wool type	Use	Fibre diameter (mµ)	Clean dry yield (%)
Fine Wool	High quality, light weight wooden clothes	17-20	65-70
Medium wool	Good quality, heavier woollen clothes	22-24	70-75
Carpet wool*	Carpets	25-32	80-90

* Carpet wool should contain a mix of wool fibre and medullated fibre (hair) in the ratio of 65:25 (w/w). Kemp fibre should not exceed 4%.

Indian Wool: In India, the process of wool shearing is generally conducted biannually with the aid of specialized shearing scissors. This task necessitates a considerable degree of expertise, making it most effectively executed by seasoned professionals on a contractual basis. Prior to shearing, sheep may be washed a few days in advance; however, this does not guarantee the removal of all contaminants, such as twigs and small stones. The act of shearing particularly soiled wool is referred to as dagging. The wool harvested from a sheep during a single shearing session is termed a fleece.

Care must be taken during shearing to prevent exposing the sheep to temperature stress. It is advisable to avoid shearing during winter, the rainy season, or during the early stages of suckling or late pregnancy. The optimal period for shearing is just before the onset of summer. Rams are typically sheared prior to mating, and lambs usually undergo their first shearing at approximately eight months of age. Common complications that may arise during shearing include cuts or injuries to the sheep and double cutting, which can lead to reduced fiber lengths. Shearing should be performed at a suitable distance from the skin, ensuring it is neither excessively close nor too far.

Within the Indian subcontinent, each geographical area is home to distinct sheep breeds. In north-western India, the Bikaneri breeds are characterized by their small size and coarse wool. Sheep from peninsular India can be classified as either coarse wool (e.g., Deccani) or hair type (e.g., Mandya), while those from the Himalayan region are predominantly wool-bearing. Most Indian breeds exhibit a low fleece yield, with the annual total yield of greasy wool typically ranging from 1 to 1.5 kg. Wool produced from tropical sheep breeds is coarse and particularly well-suited for carpet manufacturing. Consequently, India ranks among the top producers and exporters of carpet wool on a global scale. A significant volume of carpet wool is exported to developed nations. Grade designations serve to indicate the characteristics and quality of wool, as outlined in Schedules I to VI. These schedules are instrumental in categorizing wool based on its origin, extraction method, and processing stage, which is vital for quality control, trade, and effective utilization in textile production.

Table 28: Grade designations of Indian wool

Schedule	Indian wool
Schedule I	Indian Clipped Wool
Schedule II	Indian Pulled Wool
Schedule III	Indian Tannery Wool (limited) 1. Wools other than South Indian Tannery and Aden type 2. South Indian Tannery and Aden type wools
Schedule IV	Indian Mixed Wool 1. Clipped - Carded 2. Clipped - Pulled
Schedule V	Indian Hill Wool, Greasy 1. Clipped
Schedule VI	Indian Ginned Wool

Table 29: Indian wool types and their corresponding mean fibre diameters

Serial No.	Type	Mean Fiber Diameter (μ)
1	North India superior clothing white	≤ 30
2	North India clothing white	30 - 42
3	North India rug white	43 - 46
4	North India carpet type	≥ 46
5	South India blanket type	43 - 46
6	South India tannery type	≥ 46

Defects In Wool: Some common defects found in wool include hairiness, impurities, lack of uniformity, and cotts, among others.

1. **Hairiness:** The existence of wool hair within the genuine wool fleece is regarded as a detrimental factor.
2. **Breaks:** Wool is particularly susceptible to a variety of nutritional influences, diseases, and environmental fluctuations such as climatic conditions and pasture quality. Elements like drought, fever, malnutrition, and even pregnancy and lactation can result in inferior quality fibers that are more likely to break.
3. **Cotts:** This term describes a condition where coarse fibers that have shed into the fleece become entangled or felted together.
4. **Lack of uniformity:** Discrepancies in the length and diameter of fibers in neighbouring sections of the fleece are considered undesirable.
5. **Impurities:** Impurities present in the fleece, whether of vegetable or mineral origin, as well as brands created with hot irons or unwanted paint, and stains resulting from urine, parasites, plants, or bacteria, are taken into account. Certain parasitic skin diseases can also contribute to degradation.

Factors Affecting Wool Quality: Wool faults are imperfections that reduce its manufacturing value and can be classified into four primary categories.

1. Faults arising from age, genetics, and inherited defects. These are issues that can frequently be managed through genetic selection. However, advancements in this area can be gradual and necessitate meticulous selection of breeding stock.
2. Faults stemming from environmental conditions. Many of these factors are often outside the wool producer's control.
3. Faults resulting from inadequate management. Wool growers typically have the most influence over these types of faults.
4. Faults induced by infectious agents and insects.

1. **Faults arising from age, genetics, and inherited defects**
   a) **Hairy or medullated wool:** This type of wool is coarse and relatively straight, exhibiting a chalky appearance, a thick medulla (core), and a thin cortex.
   b) **Kemp:** This wool features a sharply pointed tip, and the base of the fiber ends in a bulb or brush. The fiber is short, flattened, and has a chalky appearance, with abrupt kinks in the fibers.
   c) **Black, brown, or grey pigmented fibers:** This defect arises from melanin granules generated by cells known as melanocytes. It can impact both skin and wool.
   d) **Stringy yolk:** This condition presents as strands of yellow, greasy substance (1/8 of an inch in width) that lie perpendicular to the skin. The yellow substance results from hyperactive sebaceous and sweat glands that generate excessive quantities of yolk (a mixture of oil, sweat, and salts). Typically, this surplus yolk is produced by a very localized area of skin.
   e) **Doggy wool:** This issue primarily affects older sheep and results in a lack of crimp in the wool. The affected wool maintains normal tensile strength and a lustrous appearance.
   f) **Cotted or matted wool:** Cotted wool develops when frequent wetting impacts animals with uneven fleeces, leading to the entanglement of fibers in regions such as the legs, belly, sides, and back. Cotted wools are categorized as either soft or hard, depending on the severity of the cotting. This problem can be linked to both genetic predispositions and environmental factors.
2. **Faults stemming from environmental conditions**
   a) **Non-scourable canary yellow:** This issue can occur in sheep of any age and is a result of elevated temperatures and high humidity. The discoloration typically manifests first on the belly wool before spreading to other lower areas of the fleece.
   b) **Yellow banding or fleece rot:** This condition affects the back wool of fine wool breeds, leading to the formation of yellow horizontal bands where the yolk is dry and hard, but becomes soft and pliable when wet. Poor drainage and damp skin surfaces exacerbate the yellow banding. Despite its designation, the fleece does not actually undergo rotting. This problem stems from a combination of genetic factors and environmental conditions.

c) **Weathered wool:** Wool that has been subjected to repeated wetting and excessive or intense sunlight will become brittle and lose its normal strength. These conditions can frequently lead to a reduction in normal wool production in the back areas. When dyeing weathered, damaged wool, the weathered tip will absorb dye lighter than the un-weathered butt end.

d) **Dingy wool (sandy, earthy, dusty, and muddy):** Wool that has become discoloured due to dirt and dust typically exhibits reduced strength, contains elevated levels of yolk, and therefore commands lower market prices.

e) **Seeds and burrs:** Wool contaminated with plant materials will receive a lower grade and may incur a diminished price, particularly if the burrs or seeds are challenging to eliminate.

f) **Bracken fern stain:** The bracken fern can impart a brown discoloration to wool during the grazing seasons of spring and summer. The staining caused by the fern is such that it cannot be removed through scouring, leading to significant price reductions upon sale.

g) **Log stain (charcoal):** Discoloration arises when dust or ash accumulates on sheep grazing in areas that have been burned, resulting in a grey to nearly black hue in the wool, which can be eliminated through scouring.

3. Faults resulting from inadequate management

a) **Weakness or break in wool:** This condition may result from abrupt seasonal transitions or circumstances that induce stress in the animal. Stress-inducing factors include inadequate nutrition, illness, pregnancy, parturition, lactation, and sudden exposure to cold and wet environments. These factors lead to a narrowing of the fiber, resulting in defects.

b) **Copper deficient wool (steely):** A decrease in copper levels in the diet often results in observable alterations in the wool, such as loss of crimp, diminished tensile strength, and abnormal pigmentation in coloured sheep. This condition may be confused with doggy wool; however, a significant distinction is that copper-deficient wool is weaker than its normal counterpart.

c) **Scourable diffuse yellow:** This issue results in "butter-coloured" wool and can occur in long-wooled breeds subjected to high feeding regimens prior to a show or sale.

d) **Faecal stain:** Staining of wool around the anus and hind legs can occur when faecal matter comes into contact with these areas. Carotene and chlorophyll from green plants present in the faeces can stain the wool a yellow to green hue. The price reduction associated with these wools can be mitigated through the practice of crutching.

e) **Dipping damage:** Numerous dipping solutions include remnants of colouring agents such as $CuSO_4$ or bluestone (Copper sulphate), which have the potential to discolour the wool. The use of unclean dipping vats may also result in banding or the appearance of "dip marks".

f) **Phenothiazine stain:** Phenothiazine appears as an olive-green powder or liquid suspension, frequently utilized to combat internal parasites. Negligent drenching, leaking applicators, unclean hands, or excessive salivation from sheep can lead to a permanent blue-black stain on the fleece. Additionally, the feces and urine of treated animals may impart a vivid red hue to the wool.

g) **Brands:** Brands consist of coloured substances employed to identify the wool of sheep. The presence of brand marks is regarded as a significant defect and results in a considerable reduction in price. In certain nations, the use of coloured materials on the fleece is prohibited.

h) **Skin bits:** These refer to actual fragments of skin from the sheep that remain attached to the wool fibers (locks). Such occurrences are attributed to careless, negligent, and inexperienced shearers.

i) **Second cuts:** This term describes the action of a shearer revisiting a previously sheared area to remove a very short layer of wool. The resulting short fibers lack commercial value and diminish the overall quality of the fleece.

4. Faults induced by infectious agents and insects

a) **Dermatophilosis (mycotic dermatitis or lumpy wool):** This condition is a bacterial skin infection instigated by the organism *Dermatophilus congolensis*. It can manifest at any time of the year and can affect animals of any age. Young, fine-wool breeds are particularly vulnerable. The infection results in a discharge or exudate that binds the fibers together. It is believed that the disease is transmitted through contaminated equipment or insects. The culling of infected animals is essential for managing and preventing the spread of the disease.

b) **Green and brown branded stains:** The discoloration is attributed to the bacterial organism Pseudomonas. It impacts sheep with both fine and coarse wool. This organism is present in both soil and water, thriving when the fleece remains damp for extended periods.

c) **Blue banding:** This issue is also attributed to Pseudomonas. It is a rare condition that can be scourable (removable from the wool) and is exacerbated by prolonged moisture.

d) **Black fungus tip:** This ailment is the result of a fungus that damages the final ½ inch of the fiber tip. It leads to a permanent black discoloration in wool produced under high rainfall conditions.

e) **Pink tip:** This issue arises from bacteria that thrive in cold, damp environments. They induce a red to light pink discoloration (up to 1 inch in length) in the fibers of black wool. This discoloration is scourable.

f) **Pink rot:** This condition is similarly caused by bacteria that prosper in moist environments. The bacteria lead to the deterioration of fibers (ranging from 1-2 inches). All wool breeds are susceptible. Maintaining a dry fleece with adequate ventilation is crucial to prevent this issue.

g) **Diffuse yellow or Apricot stains:** This condition frequently occurs on the lower sections of the fleece, where the belly remains persistently damp due to rainfall and prolonged exposure to wet pastures. This issue is not scourable and results in significant fiber damage.

h) **Purple stain:** Bacteria induce horizontal bands of reddish to purple discoloration in the wool. While the purple can be washed out, this process will leave the wool with a pink hue. It is vital to keep the wool dry to avert this problem.

i) **Ked stain:** The sheep ked (*Melophagus ovinus*) is a wingless parasite that feeds on the blood of animals. Although commonly referred to as sheep ticks, they are not ticks. The ked is reddish to gray, approximately ¼ inch long, and possesses six legs. This parasite moves swiftly between animals, and moderate to heavy infestations can inflict severe damage to the wool. The ked damages the wool in two primary ways. First, if the infestation is significant, the keds will extract enough blood to adversely impact the sheep's nutritional health, leading to poor wool growth. Secondly, the excrement of the keds stains the wool a reddish colour, and pupal stages of the ked may be present in the wool. The ked is also responsible for skin damage, often referred to as 'Cockle', and produces a strong odour. Effective control of this parasite can be achieved through methods such as dipping, spraying, dusting, and the application of specific pour-ons.

j) **Leg lice:** These sucking lice result in the development of a spongy, malodorous mass within the wool. The areas most commonly affected include the dew claws, shank, scrotum, and belly. These lice exhibit a blue-grey to brown coloration, possess a life cycle of approximately 43 days, and can endure off the sheep for as long as 18 days. Infections may propagate through direct contact among animals or when uninfected sheep graze on pastures that are contaminated.

Impurities of Wool: Raw wool frequently contains dirt, with nearly 50% of its weight made up of natural and other impurities. Typically, finer wools, such as merino, have a greater percentage of natural impurities in comparison to coarser wools. The types of contaminants present can differ based on various factors, including the breed of the sheep, its nutrition, the environment, and the specific location of the wool on the animal. Natural impurities encompass wax, suint, and dried sweat, while acquired impurities include dirt, grass, seeds, straw, burrs, brambles, sticks, and other plant materials. Furthermore, a layer of proteinaceous contaminants (PCL) may also be found. Additionally, during the processes of spinning and weaving, further impurities are introduced.

Table 30: Impurities in Raw Wool (in percentage)

Type	Fat and suint	Sand and dirt	Vegetable matter	Wool fibre
Fine	20-50	5-40	0.5-2	20-50
Medium	15-30	5-20	1.5	40-60
Long	5-15	5-10	0.2	60-80

The major impurities of wool consist of following

1. Wool waxes are extracted from the grease during the scouring process. These waxes consist of various monocarboxylic, dicarboxylic, and hydrocarboxylic acids, along with steroidal alcohols. Research indicates that unscoured wool contains an unoxidized portion of wool grease and other impurities that can be easily eliminated and recovered, as well as an oxidized portion located at the tip of the hair, which is challenging to remove and distinguish from other oxidized impurities.
2. Suint is typically regarded as a composition of water-soluble substances that can be easily eliminated through scouring.
3. The contaminants extracted from the scoured wool comprise both inorganic and organic substances.
4. The proteinaceous material is made up of skin flakes from the sheep and soluble peptides. Wool impurities can be categorized as natural, acquired, or applied.

   a) **Natural:** Natural impurities encompass oils and fats secreted by the sebaceous glands, as well as water-soluble salts from dried skin excretions known as suint.

   b) **Acquired:** Acquired impurities include sand, dirt, burrs, grasses, dust, and faecal matter.

c) **Applied:** Applied impurities consist of tar, paint, and other substances used for animal identification, in addition to chemicals from treatment dips and sprays, which can leave residues on the wool. These contaminants may adversely impact wool quality and potentially result in reduced market prices.

Process to Remove the Impurities: The removal of impurities from wool involves several essential steps to guarantee that the wool is clean and suitable for processing. These impurities may include lanolin (natural grease), dirt, vegetable matter (such as seeds and leaves), and other debris. The removal of wool impurities can be achieved by adhering to a specific process.

1. Pre-sorting
2. Wetting
3. Skirting
4. Crabbing
5. Scouring
6. Carbonizing
7. Milling
8. Bleaching

1. **Pre-sorting:** Inspect the wool for any visible debris, including twigs, leaves, or larger particles of dirt. Manually remove any significant contaminants. Shake or gently brush the wool to remove loose debris.
2. **Wetting:** The initial treatment applied to wool is wetting. This process releases latent strains and provides a permanent set. Wet treatment should not be conducted at temperatures exceeding those used in crabbing.
3. **Skirting:** Skirting is a crucial phase in the wool preparation process, which entails the removal of undesirable components from the fleece prior to further processing. This step involves eliminating less desirable sections of the fleece, such as contaminants, coarse wool, and other impurities, to ensure that only high-quality wool advances to subsequent processing stages. The fleece is laid out on a clean, flat surface, such as a skirting table, allowing for effective inspection and sorting. The fleece is scrutinized to identify areas with visible dirt, vegetable matter (including seeds and burrs), or other contaminants. The wool is classified according to quality and levels of contamination.
4. **Crabbing:** This treatment is applied to woollens to prevent the tendency to cockle or distort. The wool is tightly looped around a roll made of iron. This roll is a perforated cylinder covered with cotton cloth to avoid staining. It is rotated during the treatment process. The fabric is then exposed to heat, typically through hot water or steam. The temperature and duration of this treatment are meticulously controlled to ensure that the wool fibers respond appropriately without sustaining damage. Steam is introduced into the cylinder at a pressure of 40-150 lb/square inch as

needed. Subsequently, the wool is unwound and rewound, causing the outer roll of wool after crabbing to become the inner roll, and steam is applied again, enhancing the wool's affinity for dyes. The pH level of the crabbing water influences the setting of the wool; a low pH results in minimal setting, while the maximum degree of setting is achieved at a pH of 10.2.

5. **Scouring:** The scouring process for wool is distinct from that of cotton. To begin with, wool has a significantly higher content of wool grease, ranging from 30% to 60%, in contrast to cotton, which contains only about 0.5% of oil and wax. Additionally, wool is susceptible to rapid degradation when exposed to alkali, necessitating the saponification of its oils and fats to be conducted with alkali; this process must be executed with caution and at temperatures below boiling. In this context, sodium hydroxide is often substituted with sodium carbonate, ammonia, or ammonium carbonate.

Raw wool undergoes scouring through a counter current method, utilizing a machine equipped with four or five bowls arranged sequentially, allowing the wool to flow directly from the first bowl to the second, and so forth. Each bowl is fitted with a wringer at the exit, a false bottom, and rakes. Beneath the false bottom lies a spirally fluted shaft that rotates, transporting the accumulated solid dirt to a central outlet for disposal. The rakes facilitate the forward movement of the wool beneath the liquor's surface while also agitating the mixture to maintain the dirt and emulsified grease in suspension. After the wool passes through the wringer, the scouring liquor is returned to the bowl. This procedure is repeated in each bowl, culminating in a final wash with water.

Table 31: Composition of scouring liquor with soap

Bowl	Soap in solution	Sodium carbonate	Temperature
First bowl	2-3%	3-4%	49-52°C
Second bowl	1-3%	2-3%	46-49°C
Third bowl	1-1.5%	1-2%	43-46°C
Fourth bowl	Water only	0	40.5-43°C

The pH level of the washing solution must not surpass 10. Soap can interact with hard water, resulting in the formation of calcium and magnesium salts, which may reduce its efficacy. Consequently, soaps have been largely supplanted by synthetic detergents. Synthetic detergents, including Gardinol and Teepol, demonstrate greater effectiveness in hard water environments and maintain stability in acidic conditions. Furthermore, they do not get depleted during the scouring process and can be reused.

Applying soap to wet wool can cause felting if pressure is exerted, which is not desirable. Synthetic detergents alleviate this problem as they do not promote felting and are more suitable for various water conditions.

Table 32: Composition of scouring liquor with synthetic detergent

Bowl	Synthetic detergent	Sodium carbonate	Common salt	pH	Temperature
First bowl	0.25%	0.25%	0	9.0	54°C
Second bowl	0.2%	0.2%	0.4%	10-10.5	52°C
Third bowl	0.12%	0.02%	0.5%	10	49°C
Fourth bowl	0-0.1%	0	0	-	46°C

After processing every 1000 lbs (454 kg) of wool, it is essential to reinforce the scouring bowl to ensure optimal efficiency. Woven and knitted wool fabrics are frequently stitched together to create continuous ropes for the scouring process. These fabrics typically possess lower concentrations of fats and other contaminants and are treated with a 0.5% soap solution or surfactant at a temperature of 40°C. In cases where alkali is required, ammonia is the preferred choice.

6. **Carbonizing:** Wool inherently contains varying amounts of burr. If these burrs are not eliminated, they can pose significant challenges throughout the manufacturing processes. Burrs, primarily made up of cellulosic material, must be extracted from animal fibers. To achieve this, a carbonizing process is employed. This process entails treating the fabric with a 6-8% sulfuric acid (H□SO□) solution, followed by drying at temperatures between 60-70°C. Subsequently, the fabric is heated to 105-110°C and ultimately raised to 150°C to guarantee complete carbonization. After this procedure, the fabric undergoes processing through a milling machine, which effectively removes hydrocellulose and hemicelluloses, thereby eliminating all vegetable impurities from the wool.

7. **Milling:** The milling process can take place either prior to or following the dyeing procedure, contingent upon the type of wool utilized. When wool is wet and subjected to pressure, it tends to felt permanently, especially when treated with soap, alkali, or acid. This felting results in fabrics that are denser, more durable, and aesthetically pleasing. There are three primary types of milling: soap, grease, and acid.

   a) **Soap milling:** This technique employs soap to facilitate the felting process. The soap aids in cleansing the wool while promoting easier interlocking of fibers, resulting in a softer and more compact fabric. It is commonly utilized for wool garments and textiles where a desirable hand feel is sought.

b) **Grease milling:** This method utilizes natural oils or greases to aid in the felting process. By employing this technique, the durability and strength of the fabric can be enhanced while preserving some of the inherent qualities of the wool. The outcome is often a heavier and more robust fabric, making it suitable for outerwear and durable textiles.

c) **Acid milling:** This process employs acidic substances to facilitate felting. Typically, acid milling results in a fabric that is denser and firmer, as the acidic conditions promote tighter bonding among the wool fibers. This milling technique is frequently utilized to produce high-quality wool fabrics that necessitate specific textures and durability.

8. **Bleaching:** The yellowish tint present on the fabric can be eliminated only if the products are intended for sale as white or light colours, while dark colours are treated with dye. Bleaching can be accomplished through the following methods:

a) **SO_2:** This economical process, referred to as staving, involves burning sulphur in chambers where the wool is suspended in loop form on wooden poles. The sulphur generates sulphur dioxide, which interacts with the yellow pigments.

b) **Hydrogen peroxide:** Hydrogen peroxide (H□O□) is widely recognized as an effective bleaching agent for wool and other textiles, noted for its lower environmental impact compared to alternative bleaches. To enhance its bleaching effectiveness, hydrogen peroxide is typically heated to temperatures between 40-50°C (104-122°F). Solutions of hydrogen peroxide often include acids to stabilize the mixture. To neutralize these acids and maintain the pH balance, sodium silicate is incorporated. Acting as a buffer, sodium silicate ensures the stability and efficacy of the bleaching solution. The fabrics are treated without tension to prevent distortion or damage during the bleaching process. They are immersed in the hydrogen peroxide solution and left overnight to achieve complete bleaching. This prolonged contact time guarantees thorough and uniform bleaching. Following the bleaching process, the fabrics are rinsed with water to eliminate any residual hydrogen peroxide. The materials are subsequently rinsed with a diluted solution of acetic acid to counteract any residual alkalinity from the sodium silicate or other substances. Ultimately, a comprehensive wash with water is performed to eliminate any leftover acetic acid or contaminants.

17

Processing of Wool

The processing of wool encompasses multiple stages that convert raw fleece into functional fabric or yarn. Each phase in the wool processing procedure is vital in influencing the ultimate quality and appropriateness of the wool product for diverse uses. Below is a concise overview of the steps involved in wool processing:

1. Shearing or pulling
2. Skirting
3. Washing or scouring
4. Carding
5. Combing
6. Spinning
7. Dyeing
8. Weaving and knitting
9. Finishing
10. Packaging

1. **Shearing or pulling:** Shearing refers to the procedure of extracting fleece from a sheep. This process is performed to gather wool from the sheep, which is subsequently utilized in the production of textiles and various other items. Typically conducted once annually during the spring or early summer, shearing occurs after the wool has reached full development, yet prior to the onset of summer heat. In certain areas, shearing may take place in the fall or winter, contingent upon local climatic conditions and the specific breeds of sheep. It is essential that shearing occurs in a clean, dry environment to reduce the risk of wool contamination. Specialized shearing clippers or electric shears are employed for this task, while hand shears have become less prevalent due to their lower efficiency. The sheep is generally positioned on its side or back, depending on the shearer's preference and the comfort of the sheep. The fleece is extracted in a single piece or in substantial sections, with the objective of minimizing cuts and bruises on the sheep

while ensuring a clean and complete fleece. Following shearing, the fleece is handled with care to prevent contamination or damage, often being rolled or neatly folded.

Pulled wool is obtained from the skins of slaughtered or deceased animals through one of several pulling techniques:

- **Sweating:** This method entails placing the skins in a humid chamber at 20°C (68°F) for a duration of 48 hours, which induces controlled decomposition that loosens the hair follicles.
- **Painting:** A dehairing agent, such as sodium sulphide, is applied to the skin to effectively loosen the hair follicles.
- **Liming:** The pelt is immersed in a limewater solution, which aids in detaching the hair from the skin.

Subsequently, the fleece is skirted to eliminate any unwanted components, such as dirt or plant matter, prior to processing. The raw fleece is stored in a dry, clean environment to avert spoilage before it undergoes further processing. The fleece is sorted according to quality, taking into account factors such as fibre length, diameter, and colour, before proceeding with additional processing.

2. **Skirting:** Skirting is a crucial phase in the wool preparation process that entails the elimination of undesirable elements from the fleece prior to further processing. Skirting pertains to the extraction of less desirable components of the fleece, including contaminants, coarse wool, and various impurities, to guarantee that only high-quality wool advances to subsequent processing stages. The fleece is laid out on a clean, flat surface, such as a skirting table, allowing for effective inspection and sorting. The fleece is scrutinized to detect areas with visible dirt, vegetable matter (like seeds and burrs), or other contaminants. The wool is classified according to quality and levels of contamination. Certain sections may be considered unsuitable for higher-quality products. Portions with significant contamination from dirt or dung are excised and discarded. Wool that contains considerable plant material is removed to prevent compromising the quality of the final product. Coarse fibers or kemps that fail to meet quality standards are eliminated, as these fibers can influence the texture and uniformity of the processed wool. Following the removal of undesirable components, the remaining wool is sorted by quality, often into various grades or categories, such as fine wool or coarser wool. The sorted wool is prepared for further processing, ensuring it is cleaner and more uniform. Skirting enhances the consistency of the wool by eliminating inconsistent or lower-quality fibers and guarantees that the wool is appropriate for specific applications,

such as fine textiles or carpets. The removal of contaminants and lower-quality sections diminishes the amount of cleaning and processing needed later, thereby saving time and costs. Higher-quality wool commands better prices in the market, making skirting instrumental in achieving improved economic returns. The skirted wool is stored appropriately to preserve its cleanliness and quality before further processing. Wool may be labeled or categorized according to the skirting process to promote efficient processing. Even after the skirting process, wool may require further cleaning or sorting prior to spinning or other processing phases. Skirting plays a vital role in the preparation of wool, ensuring that it is clean, of high quality, and adequately prepared for the next stages of processing. Effective skirting is essential for preserving the integrity and value of the wool, which benefits both producers and end-users.

3. **Washing or scouring:** The washing or scouring process is a fundamental aspect of wool processing that entails cleaning raw wool to eliminate impurities such as grease (lanolin), dirt, and vegetable matter. Scouring specifically refers to the washing of raw wool to remove natural grease, dirt, and other contaminants. This step is critical for preparing wool for subsequent processing and enhancing its quality. Scouring is generally conducted using specialized washing machines or scouring facilities that utilize hot water and detergents. In smaller operations, this may be performed manually. Hot water is employed to aid in dissolving grease and other impurities, with temperatures typically ranging from 40°C to 70°C (104°F to 158°F). Initially, the wool is soaked in hot water to loosen and soften the grease and dirt. Gentle agitation or stirring is applied to dislodge the impurities from the wool fibers. Specially formulated wool detergents or soaps are introduced into the water, designed to break down lanolin (wool grease) and emulsify dirt. Occasionally, conditioners are incorporated to help preserve the wool's natural characteristics and prevent damage during the washing process. The wool is then rinsed with clean hot water to eliminate the detergent, grease, and any remaining impurities. Multiple rinsing stages may be necessary to ensure the complete removal of all detergent and contaminants. Following rinsing, excess water is drained from the wool. In certain processes, centrifugation is employed to extract additional water and expedite drying. The cleaned wool is typically spread out in a clean, well-ventilated space to air dry. In industrial environments, controlled drying techniques may be utilized to guarantee that the wool dries evenly and efficiently. Eliminates lanolin, dirt, and other contaminants that may compromise the quality and appearance of the

final product. Cleaner wool facilitates processing and yields superior yarn or fabric. Clean wool is simpler to card and comb, resulting in a more consistent and higher-quality product. For optimal dye absorption and vibrant colours, wool must be clean. Effective scouring lowers the costs of subsequent processing by reducing the necessity for additional cleaning phases. Adopting proper scouring techniques aids in managing wastewater and lessening environmental impact through responsible treatment and disposal of contaminants. Scouring is an essential phase in converting raw wool into high-quality fiber, ensuring it is clean and prepared for the next processing stages. Adhering to proper scouring methods enhances the wool's applicability for diverse textile uses and contributes to the overall efficiency and effectiveness of wool processing.

4. **Carding:** Carding represents a crucial phase in the wool processing sequence that converts cleaned wool into a form suitable for spinning into yarn. This process entails disentangling, separating, and aligning wool fibers to ready them for further processing. Carding involves breaking up and aligning wool fibers to form a continuous web or batt of fiber. This stage assists in eliminating any remaining impurities and prepares the wool for spinning. Prior to carding, the wool must be clean and devoid of excess moisture, typically following scouring, where the wool is washed and dried. Carding is executed using carding machines, which comprise a series of rotating drums equipped with carding teeth or wires. Clean wool is introduced into the carding machine, usually in the form of loose fleece or batts. The machine drafts (stretches) the wool fibers to initiate the separation and alignment process. The carding drums, fitted with fine wire teeth, pull the wool apart and segregate the individual fibers. The wool fibers are aligned parallel to one another, resulting in a more uniform and manageable structure. The carded wool is transformed into a continuous web or batt, which is a soft, mat-like layer of aligned fibers. In certain instances, carding machines yield a sliver, a continuous strand of carded wool fibers that can undergo further processing. Carding serves to eliminate short fibers, neps (small knots), and any residual impurities. The outcome is cleaner, more uniform wool that is simpler to spin. Hand carders are comb-like instruments utilized for smaller quantities of wool. They comprise two paddles equipped with fine wire teeth. Wool is manually carded by brushing it back and forth between the paddles, aligning the fibers and forming a batt. Machine carding is more efficient and appropriate for high-volume processing, generating consistent batts or slivers. These machines are employed for large-scale wool processing and consist of multiple

rotating drums and rollers. Carding aligns wool fibers, facilitating their spinning into yarn, and produces a more uniform fiber preparation, which enhances the quality of the yarn and minimizes inconsistencies. Carding aids in the removal of residual debris, short fibers, and neps that could compromise the quality of the final product. Carded wool is softer and more uniform, which is crucial for the production of high-quality textiles. Over-processing or employing excessively aggressive carding settings can harm wool fibers, impacting their quality. Proper maintenance and settings of the machine are essential to ensure efficient and effective carding without excessive waste. Carded wool may be subjected to combing to further align fibers and eliminate short or coarse fibers, especially for finer yarns. The carded wool is then spun into yarn, which can subsequently be dyed and woven or knitted into fabric. Carding is a vital process in wool preparation, ensuring that wool fibers are adequately aligned and prepared for the subsequent stages of spinning and textile production. It plays a significant role in determining the quality and texture of the final wool products.

5. **Combing:** Combing represents an essential phase in the processing of wool, further enhancing carded wool to yield a superior product. This process entails the alignment and separation of wool fibres, effectively eliminating shorter fibres and impurities, which culminates in a smoother and more uniform wool top or sliver. The outcome of this step is a refined wool top or sliver, ideally suited for the spinning of fine yarn. Initially, the wool designated for combing is carded, resulting in a fluffy, untangled batt or sliver. The combing process employs specialized machines or hand tools, such as combs or combing cylinders. The carded wool is introduced into the combing apparatus or combs, typically in the form of a sliver or batt. Combing machines utilize a series of rotating combs or cylinders equipped with fine, closely spaced teeth to separate and align the wool fibres. During this procedure, shorter fibres (noil) and impurities are extracted. These shorter fibres are generally deemed undesirable for fine yarns, as they may introduce inconsistencies. The outcome of combing is a smooth, continuous wool top, representing a high-quality preparation of aligned, long fibres. Alternatively, a continuous sliver of combed wool is generated, which can be directly spun into yarn. For small quantities of wool, hand combs or brushes are employed to manually comb the fibres. This manual method is typically applied to fine or specialty wools, allowing for the alignment of fibres and the removal of impurities in small batches. In contrast, combing machines are utilized for large-scale wool processing,

comprising multiple rotating combs or cylinders that automate the combing procedure. Machine combing is characterized by its efficiency and suitability for high-volume production, yielding consistent wool tops or slivers. The alignment of wool fibres in parallel enhances the consistency and smoothness of the final yarn, resulting in a higher-quality wool preparation that is appropriate for fine textiles and premium products. Furthermore, combing effectively removes shorter fibres that could compromise the strength and texture of the yarn. Short fibres, also known as noils, are eliminated during the combing process to yield a cleaner and more refined wool top or sliver. The combing of wool results in smoother and stronger yarns, which are essential for the production of high-quality fabrics and textiles. This process effectively removes short fibres and noils, leading to some waste; however, this waste can be repurposed for lower-quality products. Combing is a vital stage in the refinement of wool for high-quality textile manufacturing, ensuring that the fibres are well-aligned, clean, and prepared for spinning into superior yarns.

6. **Spinning:** Spinning represents a crucial process in the processing of wool and fibres, converting prepared wool or other fibres into yarn or thread. This procedure entails drawing out the fibres and twisting them together to form a continuous strand. The spinning process transforms prepared wool or other fibres into yarn by elongating and twisting the fibres together, resulting in a continuous strand of yarn that is suitable for weaving, knitting, or other textile applications. Prior to spinning, fibres undergo carding and combing to ensure they are aligned and clean. The prepared wool fibre is generally presented in the form of a sliver or roving. Spinning can be executed using a variety of equipment, ranging from traditional spinning wheels to contemporary spinning machines. The sliver or roving is drawn out to lengthen and thin the fibres, which aids in aligning them and reducing their thickness. The objective is to achieve a uniform yarn thickness by manipulating the tension and draft of the fibres. The drawn-out fibres are then twisted together to form a continuous strand of yarn. The degree of twist influences the yarn's strength, texture, and elasticity. Different techniques can be employed to produce various types of yarn, such as single-ply or multiple-ply yarns. The spun yarn is subsequently wound onto bobbins or spools, preparing it for dyeing, finishing, and further textile production. In industrial spinning, yarn is wound onto bobbins or cones, making it ready for use in weaving or knitting. Plying is the process of twisting together two or more strands of yarn to form a thicker and more robust

yarn. This procedure can enhance both the texture and strength of the yarn. A variety of plying methods are employed to achieve different effects, including cable ply and spiral ply. Traditional hand spinning typically utilizes a spinning wheel, which is manually operated to draft and twist the fibers. Another conventional tool is the drop spindle, a straightforward device designed for spinning small quantities of yarn. Ring Spinning represents a prevalent industrial technique where yarn is drawn out and twisted by a ring-spinning machine, renowned for its ability to produce fine and durable yarns. Open-end spinning, a more contemporary method, employs a rotor for yarn spinning, offering increased speed and efficiency, albeit potentially resulting in a different texture than that produced by ring spinning. Air-jet spinning is yet another technique that utilizes air pressure to spin the yarn, yielding a distinctive texture and facilitating high-speed production. The spinning process is critical in determining the strength and durability of the yarn, which in turn influences the quality of the final fabric. Effective spinning results in yarn that possesses a uniform texture and appearance, which is vital for the production of high-quality textiles. Spinning enables the creation of a diverse array of yarn types, ranging from fine and delicate to thick and textured, catering to various textile applications. Spinning methods can be tailored to yield yarns with specific characteristics, such as elasticity or softness. Following the spinning process, yarn may undergo further treatments such as dyeing, setting, or finishing to improve its properties and prepare it for textile manufacturing. Spinning is an essential procedure in textile production, transforming raw fibers into yarn that is indispensable for the creation of a wide variety of fabrics and textile goods. The use of appropriate spinning techniques and equipment is crucial for generating high-quality yarn that satisfies the requirements of diverse applications.

7. **Dyeing:** The process of dyeing involves the addition of colour to textiles or fibers through the application of dyes. This stage is essential in textile manufacturing and can greatly influence the appearance, quality, and value of the finished product. Dyeing refers to the method of applying colour to fibers, yarns, or fabrics. It encompasses the use of dyes to alter the color of the material, which can enhance visual appeal, ensure specific colour fastness, and fulfil design specifications. Prior to dyeing, the fiber, yarn, or fabric must be cleaned to eliminate any impurities, finishes, or residues that could hinder dye absorption. This cleaning process may involve washing, scouring, and occasionally bleaching. The selection of dye is contingent upon the type of fiber, the intended colour,

and the final application of the textile. Common types of dyes include acid dyes, reactive dyes, and disperse dyes. The wool industry typically employs common dyeing methods such as batch dyeing, continuous dyeing, and printing.

a) **Batch dyeing:** This method entails dyeing a specific batch of fiber, yarn, or fabric within a single dye bath. Various techniques can be utilized:
   - **Piece dyeing:** This technique involves dyeing fabric pieces after they have been woven or knitted.
   - **Skein dyeing:** This method is used for dyeing yarns that are wound into skeins or hanks.
   - **Top dyeing:** This process involves dyeing wool tops prior to spinning them into yarn.

b) **Continuous dyeing:** In this approach, the fabric or fiber is continuously passed through a dye bath and processed, which is typically employed in large-scale operations. Techniques include:
   - **Pad-dry-cure:** This method applies dye to fabric using a padding technique, followed by drying and curing.
   - **Jet dyeing:** This technique utilizes a jet dyeing machine where the fabric is circulated within a dye bath.

c) **Printing:** This process consists of applying dye to specific sections of fabric to produce patterns or designs. Techniques include:
   - **Screen printing:** This method employs screens to apply dye onto fabric.
   - **Digital printing:** This technique utilizes digital technology to print designs directly onto fabric.

The dye is dissolved in water to form a dye bath. The fibre, yarn, or fabric is submerged in this bath, allowing the dye molecules to infiltrate and adhere to the material. Conditions for dyeing, including temperature, pH, and duration, are meticulously regulated to achieve the intended colour and ensure effective dye absorption. Following the dyeing process, the material is rinsed to eliminate any surplus dye and avert colour bleeding. Certain dyes necessitate fixation methods, such as heat-setting or chemical fixing, to guarantee that the dye bonds permanently with the fibre. Additional finishing techniques may be employed to improve the fabric's appearance, texture, and functionality. The dyed material undergoes testing for colour fastness, which assesses its resistance to washing, light exposure, and abrasion. It is essential to ensure uniform colour and quality across production batches to uphold product standards.

Table 33: Different dyes with uses and properties

Dyes	Uses and properties
Acid dyes	Mainly used for protein fibres like wool and silk. Provide bright, vibrant colours and good colour fastness.
Reactive dyes	Suitable for cellulose fibres such as cotton and rayon. Form a chemical bond with the fibre, offering excellent colour fastness and brightness.
Disperse dyes	Used for synthetic fibres like polyester. Designed to be dispersed in water and absorbed by synthetic fibres.
Direct dyes	Applied directly to cellulose fibres. Generally used for dyeing cotton and paper products.
Vat dyes	Often used for dyeing cotton and other cellulose fibres. Require a reduction process to make the dye soluble, which then reacts with the fibre.

Dyeing facilitates a diverse array of colours and patterns, thereby enhancing the aesthetic appeal of textiles and enabling innovative designs. Dyeing permits the creation of bespoke and distinctive textile products that cater to specific market needs. Effective dyeing guarantees that colours remain vivid and resistant to fading, laundering, and various environmental influences. Completed dyed textiles undergo inspection for colour uniformity, quality, and any imperfections. Assessments for colour fastness and durability are performed to ensure that the final product adheres to industry standards. Supplementary treatments such as softening, sizing, or setting may be applied to improve the properties of the final product. Dyeing processes produce wastewater and chemical byproducts, which necessitate management and treatment to mitigate environmental repercussions. Dyeing represents a crucial phase in the textile industry that affects the appearance, quality, and market value of textiles. Efficient dyeing processes and methodologies are vital for the production of high-quality, visually attractive, and durable textile goods.

8. **Weaving and knitting:** Weaving and knitting are two essential methods for fabric creation from yarn or thread. Each technique possesses unique processes and yields different types of textiles with distinct characteristics. Weaving is a method of textile production that interlaces two sets of yarns, namely warp (longitudinal) and weft (transverse), to fabricate fabric. The warp yarns extend lengthwise, while the weft yarns traverse crosswise, interlacing in a systematic pattern to form the fabric. Yarns are prepared and wound onto spools or bobbins. The warp yarns are positioned on the loom, and the weft yarns are arranged for insertion. The warp yarns are aligned on the loom's warp beam in parallel rows. The loom generates a shed, or opening, in the warp yarns by elevating and lowering alternate warp threads. This gap permits the weft yarn to traverse through. The weft yarn is introduced through the

shed utilizing a shuttle or other insertion devices. This operation can be executed manually or through automated means. The weft yarn is positioned against the pre-existing woven fabric with the aid of the reed or batten, which guarantees that the fabric remains tight and compact. As additional weft yarn is woven, the newly created fabric is rolled onto the fabric beam, causing the warp yarns to advance forward. The woven fabric then undergoes a series of finishing processes, including washing, dyeing, and setting, to improve its texture, color, and overall performance. Weaving results in a diverse array of fabric structures and patterns, ranging from simple to intricate designs. This process produces robust and stable fabrics that are suitable for a variety of applications, including clothing and upholstery.

Table 34: Different types of weaving

Weaving	Characteristics	Examples
Plain weave	Simple, balanced weave with a basic over-under pattern. Produces a sturdy and versatile fabric.	Cotton fabric, muslin
Twill weave	Diagonal pattern created by weaving the weft yarns over and under multiple warp yarns. Provides durability and drape.	Denim, gabardine
Satin weave	High sheen and smooth surface achieved by weaving the weft yarns over several warp yarns, minimizing the interlacing.	Satin, charmeuse
Jacquard weave	Complex patterns and designs created using a Jacquard loom, which controls individual warp threads.	Brocade, damask
Tapestry weave	Artistic and decorative weaving where weft yarns are used to create intricate patterns and images.	Tapestries, wall hangings

Knitting is a process in textiles where yarn is looped and interlocked to produce fabric. In contrast to weaving, which involves the interlacing of two sets of yarn, knitting employs a single strand of yarn to create interconnected loops. The yarn is prepared and wound into either balls or skeins. Knitting needles or machines are arranged for the knitting procedure. The initial row of loops is established by casting on stitches onto the needle, which serves as the foundation for the knitted fabric. Stitches are formed by creating loops of yarn through the previously established stitches, utilizing either needles or knitting machines. Different types of stitches, including knit, purl, and rib, yield various textures and patterns. Knitting facilitates easy shaping and modifications, such as increasing or decreasing stitches to accommodate body contours or to achieve specific designs. The concluding row of stitches is bound off to secure the edges and avert unravelling. The knitted fabric may undergo washing, blocking, or steaming to establish its shape and enhance its texture. Knitted fabrics exhibit a high degree of flexibility and stretch, rendering them suitable for garments that must conform to body shapes. Additionally, knitted fabrics are typically softer and more comfortable against the skin in comparison to woven fabrics. However, knitted fabrics can be susceptible to snagging and unravelling if not handled or finished correctly. The creation of intricate patterns and textures necessitates skill and may pose greater challenges in hand knitting.

Table 35: Different types of knitting

Knitting	Characteristics	Examples
Weft knitting	Yarns are knitted horizontally across the fabric. Typically involves single or double knitting.	Jersey knit, rib knit
Warp knitting	Yarns are knitted vertically along the length of the fabric, creating more stable and less stretchable fabric.	Tricot, raschel
Hand knitting	Knitting is done manually using needles, allowing for customization and intricate designs.	Handmade sweaters, scarves.
Machine knitting	Uses knitting machines to automate the process, suitable for mass production.	Factory-produced knitwear, hosiery

9. **Finishing:** The finishing process represents the concluding phase in textile production, occurring subsequent to weaving or knitting, aimed at enhancing both the properties and aesthetic appeal of the fabric. This phase encompasses a variety of treatments designed to elevate the fabric's visual appeal, tactile quality, performance, and durability. Finishing pertains to the assortment of processes applied to fabric or yarn post-weaving or knitting to attain specific attributes such as texture, colour, and functionality. Its purpose is to refine the fabric and ready it for its intended application. The finishing processes enhance the fabric's visual characteristics by introducing gloss, texture, or colour. They also improve the tactile experience of the fabric, rendering it softer, smoother, or more textured according to requirements. Treatments such as water repellency and flame retardancy augment the functional properties and longevity of the fabric. Additionally, softening and breathable finishes contribute to the comfort and usability for the wearer. Finishing can bestow particular properties essential for specific applications, including UV protection for outdoor textiles or stain resistance for upholstery. Finished fabrics are subsequently packaged and prepared for distribution to manufacturers or retailers for incorporation into garments, home textiles, or other products. Thus, finishing constitutes a vital stage in textile production that enhances the fabric's appearance, performance, and usability. The diverse finishing processes employed can profoundly influence the final product, rendering it suitable for a wide array of applications and enhancing its overall quality.

Table 36: Different types of finishing processes

Type of processing	Processes	Description	Purpose
Mechanical finishing	Calendering	Passing fabric through a series of rollers to smooth, compress, and impart a high sheen.	Creates a smooth, glossy surface and can also increase fabric density and weight.
	Brushing	Fabric is brushed with rotating wire brushes to raise the fibres and create a soft, fuzzy texture.	Enhances softness and can create a nap on fabrics such as fleece or flannel.
	Merging	Using heat and pressure to bond fibres or layers of fabric together.	Creates a fabric with improved strength and stability, often used for bonded or laminated textiles.
Chemical finishing	Scouring	Cleaning the fabric to remove natural oils, grease, and impurities using alkaline solutions.	Prepares the fabric for subsequent dyeing and printing processes.
	Bleaching	Applying chemicals to remove natural colour and whiten the fabric.	Provides a uniform base for dyeing or printing and enhances fabric brightness.
	Dyeing and printing	Adding colour to the fabric using dyes or pigments through various methods such as batch dyeing, continuous dyeing, or screen printing.	Imparts colour and patterns to the fabric, meeting design and aesthetic requirements.
	Water repellent and water-resistant finishes	Applying treatments that make the fabric repel water or resist water penetration.	Enhances the fabric's performance in outdoor or weather-sensitive applications.
	Flame retardant finishes	Applying chemicals to make the fabric resistant to ignition or flame spread.	Improves safety for fabrics used in environments where fire resistance is required.
	Anti-microbial and anti-bacterial finishes	Treating fabric with agents that inhibit microbial growth or bacterial odours.	Enhances hygiene and longevity, especially in textiles for medical or sports applications.
	Softening	Using chemicals or mechanical methods to improve fabric softness.	Increases comfort and enhances the tactile feel of the fabric.

Physical finishing	Sanforising	Pre-shrinking the fabric by subjecting it to controlled heat and moisture.	Reduces fabric shrinkage after washing and ensures dimensional stability.
	Heat setting	Applying heat to stabilize synthetic fibres and prevent unwanted shrinkage or distortion.	Ensures fabric retains its shape and dimensions after use and washing.
	Stiffening	Applying substances to increase fabric stiffness or rigidity.	Provides structural support for fabrics used in applications like collars or upholstery.

10. **Packaging:** In the textile sector, packaging represents a vital process that guarantees the safeguarding, preservation, and effective distribution of fabrics and garments. Adequate packaging not only upholds the product's quality but also streamlines handling, storage, and transportation. It mitigates damage from physical elements such as abrasion, crushing, or exposure to environmental conditions during transit and storage. Furthermore, packaging protects against dirt, dust, moisture, and other pollutants that could compromise the fabric's integrity. It maintains the textile in prime condition, preserving its colour, texture, and overall aesthetic. Packaging ensures that the fabric or garment retains its shape and facilitates convenient handling, stacking, and storage in warehouses and during transport. It aids in organizing and managing inventory through clear labelling and tracking. Additionally, it provides a platform for showcasing brand logos, information, and promotional messages. Visually appealing packaging can enhance the product's attractiveness and sway purchasing decisions. It is essential to ensure that textiles are clean, dry, and devoid of defects prior to packaging. Fabrics or garments should be folded or rolled to fit the packaging efficiently and reduce wrinkles. Each garment or roll of fabric must be enclosed in protective covers or bags. All packages should be securely sealed using suitable tapes or adhesives. Labels containing product details, care instructions, and other pertinent information should be attached. A final inspection should be conducted to confirm that packaging is executed correctly and that all items are appropriately labelled and safeguarded. If necessary, tests should be performed to verify that packaging materials offer sufficient protection. Packaging is a fundamental process in the textile industry that ensures products are safeguarded, preserved, and distributed efficiently. It plays a crucial role in maintaining product quality, enhancing brand identity, and facilitating seamless logistics operations. Effectively implemented packaging can also elevate customer satisfaction and contribute to overall business success.

Table 37: Different types of packaging

Types of packaging	Packaging forms	Purpose
Retail Packaging	Individual packaging	For garments sold directly to consumers, often includes branded polybags or hangtags.
	Display packaging	Includes packaging designed for display on retail shelves, which may include clear windows or attractive designs.
Bulk Packaging	Cartons and pallets	Used for shipping large quantities of textiles to distributors or retailers, typically in corrugated boxes or on pallets.
	Bales	For large quantities of fabric, often wrapped in plastic and secured with strapping.
Specialty Packaging	Protective packaging	For delicate or high-value items, using bubble wrap, foam, or custom inserts for added protection.
	Eco-friendly packaging	Using recyclable or biodegradable materials to minimize environmental impact.

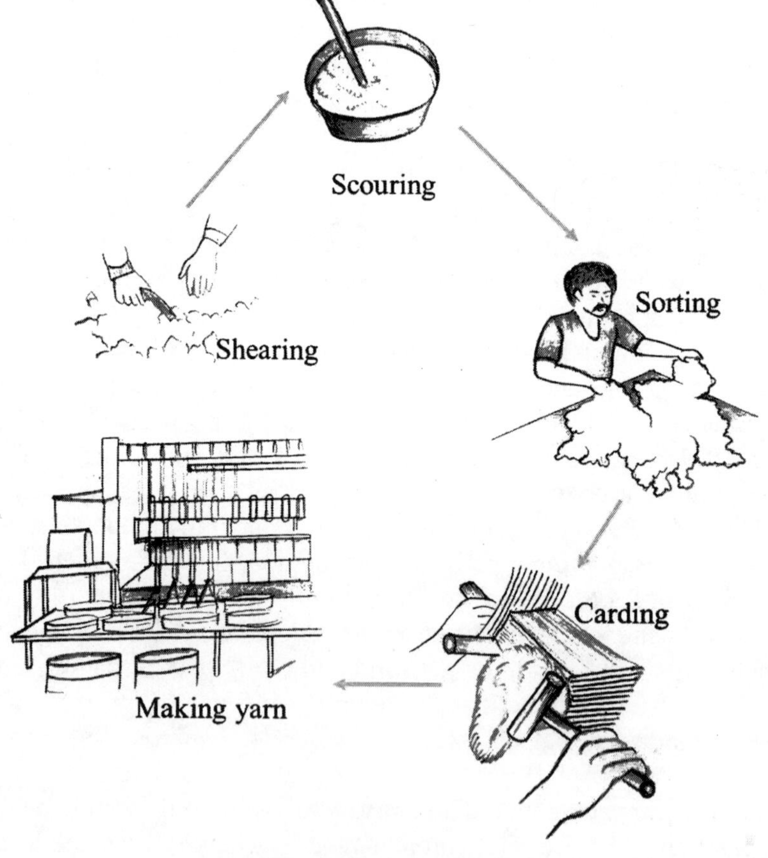

Figure 45: Processing of wool

18

Utilization of Animal Bristle

Animal bristles are natural fibers derived from the hair or fur of various animals, mainly pigs and horses. Celebrated for their distinctive characteristics, these bristles have been employed for centuries across a range of uses. Bristles consist of stiff, wiry hairs from pigs, hogs, or boars, and are typically utilized in the manufacture of different types of brushes. They are predominantly harvested from the back, neck, and tail, as the hairs located on the flank and belly are too short for effective use. Pig bristles are characterized by their coarse, stiff texture and tapered shape, often featuring split or flagged tips. These flagged tips improve their efficiency in paint and varnish applications due to their capacity to retain paint effectively.

Shaving brushes crafted from soft bristles surpass synthetic options owing to their enhanced water absorption and retention abilities. In our nation, bristles are primarily obtained from local and crossbreed pigs. Indian bristles are noted for their coarseness and durability, although their longer flagged tips (which can constitute up to 30% of their total length) necessitate trimming prior to use, potentially diminishing their overall value.

Bristles are harvested from live pigs or boars once or twice annually, as well as from slaughtered or deceased animals. In certain bacon processing facilities, bristles are also collected by shaving the skins of slaughtered animals post-scalding. Those obtained from live pigs generally exhibit superior lustre and resilience compared to those from deceased animals.

Characteristics of Bristle: The colour of bristles is a significant attribute, categorizing them into three groups: white, black, or grey. Any bristles that do not fall entirely into the black or white categories are classified as grey. White bristles are predominantly found in western Uttar Pradesh and Punjab, while grey is the most prevalent colour for Indian bristles. Another important characteristic of bristles is their length; longer bristles typically command higher prices. As the length of bristles increases, their thickness generally also rises, leading to enhanced stiffness. Bristles that are shorter than 44 mm are classified as short and rifling, whereas those measuring between 44 mm and over 159 mm are divided into 19 distinct grades as per the Bristle Grading and Making (Amendment) Rules, 1973. In India, due to the tropical climate, bristles typically range in length from 57 mm to 159 mm.

Bristles obtained from wild boars in Punjab and Madhya Pradesh are recognized for their favourable length, while high-quality black bristles, known as "Darjeeling bristles," are sourced from wild boars inhabiting the foothills of the Himalayas. These bristles are comparable to the Chinese variety, with Darjeeling being the primary market for them.

Regarding thickness, bristles are commercially categorized into three groups: Extra stiff, Stiff/Semi-stiff, and Soft. Extra stiff bristles, which are derived from wild boars, are thicker and stiffer than their stiff or semi-stiff counterparts. Conversely, soft bristles are thinner than stiff bristles. Among these categories, white, soft, and longer bristles tend to fetch the highest prices.

1. Physical properties

a) **Stiffness and flexibility:** The rigidity of animal bristles varies based on their origin. Hog bristles are recognized for their firmness, making them suitable for uses that necessitate strong pressure, such as paintbrushes designed for oil-based paints. In contrast, horsehair is softer and more pliable, rendering it more appropriate for delicate tasks, such as makeup brushes.

b) **Length and uniformity:** Bristles obtained from various animals differ in length and thickness. High-quality bristles are usually long and uniform, which is particularly crucial for precision instruments, such as fine artist brushes.

c) **Absorbency:** Natural bristles possess excellent absorbent characteristics, enabling them to effectively retain liquids. This quality makes them ideal for uses involving paints, varnishes, and cosmetics.

d) **Durability:** Although natural bristles are typically robust, they may be prone to wear and shedding as time progresses. Their effectiveness can also be affected by environmental conditions, including humidity and temperature.

2. Performance attributes

a) **Paint application:** Hog bristles are particularly proficient in applying oil-based paints due to their ability to evenly hold and release paint. Their resilience enables them to endure harsh chemicals present in certain paints without deterioration.

b) **Cosmetic use:** Animal hair utilized in cosmetic brushes provides a gentle and smooth makeup application, especially for powder products. The natural taper of the bristles facilitates blending, leading to a seamless finish.

c) **Cleaning tools:** The firmness and durability of specific animal bristles render them suitable for cleaning brushes employed in both industrial and domestic environments. Their robustness allows them to effectively address dirt and grime.

Collection of Animal Bristles

1. Sources

a) **Pigs:** Hog bristles are obtained from the back or neck regions of pigs and are extensively utilized in brush production. The quality of these bristles can vary considerably based on the particular area of the pig from which they are sourced.

b) **Horses:** Horsehair is harvested from different parts of the horse's body, with tail hair being especially valued for its length and flexibility. This characteristic makes it particularly suitable for uses that necessitate a gentler touch, such as in fine brushes.

2. Collection methods

a) **Grooming and shearing:** Hog bristles are often collected during regular grooming or shearing processes. This procedure is typically conducted in a controlled setting to ensure that the bristles are harvested without inflicting harm on the pigs.

b) **Slaughterhouse by-products:** In many cases, bristles are collected as by-products of the meat industry. This approach guarantees that the bristles are sourced ethically, reducing waste and utilizing the entire animal.

c) **Tail clipping:** The collection of horsehair is typically achieved by clipping the tail hair of horses. This procedure is conducted with great care to prevent any harm to the animal while facilitating the acquisition of high-quality hair.

Post Harvest Handling: Prior to packing, raw bristles must undergo a dressing process. This begins with soaking the bristles in warm water combined with washing soda for several hours. Following this, they are scoured with detergent, thoroughly rinsed in cold water, and dried. It is crucial to ensure that the root and flag ends remain separate throughout this process, after which the dried bristles are bundled together.

Subsequently, the dried bristles are categorized into grades based on specific lengths using a method known as "dragging." In this technique, approximately

500 grams of bristles are secured with a strap on a small wooden platform, oriented with their root ends facing downward. The longest bristles are extracted first, followed by those of progressively shorter lengths. Each size is then tied into bundles weighing around 100 grams. These bundles undergo a final "solid dressing" to guarantee uniformity in both length and size. Kanpur is recognized as the largest dressing centre in India, with Jabalpur, Allahabad, and Gorakhpur also serving as notable centres.

Bristles designated for shaving brushes or export must be sterilized to eradicate any potential anthrax spores and bacteria. This sterilization is accomplished by autoclaving at a pressure of 25-40 lbs for approximately 1.5 hours, followed by drying in a hot air oven at 60°C to restore their stiffness.

Bundles of bristles that share the same thickness and grade lengths are tied together. Each bundle, with the exception of short and rifling bristles, has a diameter ranging from 1.5 to 2 inches. These bundles are packed in clean, dry wooden cases, ensuring a net weight of 10 kg or more, in increments of 2 kg, with a maximum weight of 46 kg for shipment. For air freight, tin or aluminum cases may be utilized. Bristles with grade lengths of 121 mm or less are packed separately from those exceeding 121 mm. The wooden cases and various containers are lined with waterproof paper and are filled with insecticides, such as naphthalene balls, to safeguard the contents.

Utilization of Animal Bristles

a) **Paint brushes:** Hog bristles are extensively utilized in paintbrushes for the application of oil-based paints due to their rigidity and superior paint-holding capacity, which enables a smooth and uniform application. Brushes crafted from hog bristles are recognized for their durability and resistance to wear, even with regular use.

b) **Cosmetic brushes:** Animal hair, particularly from squirrels and goats, is employed in premium cosmetic brushes. These brushes are valued for their softness and effectiveness in blending makeup. The natural taper and flexibility of animal hair aid in the precise application and blending of makeup products.

c) **Cleaning brushes:** Stiff bristles are typically found in cleaning brushes used for both industrial and domestic cleaning tasks. Their strength and rigidity render them effective for scrubbing and eliminating debris. To ensure durability and efficiency, brushes with animal bristles necessitate proper maintenance, including routine cleaning and suitable storage.

Value Added Products from Pig Bristle: The Northeast Himalayan (NEH) region of India is home to approximately 28% of the nation's total pig population. It is estimated that around 1.535 million pigs are slaughtered each year in the organized sector. On average, each indigenous pig yields 300-400 grams of high-quality bristles, resulting in an annual production of roughly 10-12 thousand quintals of pig bristles in the NEH region.

The Indian Council of Agricultural Research (ICAR-NEH) has devised a novel methodology for collecting bristles through the clipping of live pigs, which does not impede their growth performance. The production of quality bristles can significantly boost the income of smallholder pig producers.

An internal processing method has been developed to eliminate dirt (including epithelial scales and wax) and to eradicate microbes and parasitic eggs. Prior to the drying process, bleaching is conducted to soften the bristles and remove any coloration. After processing, a range of products is manufactured from local pig bristles, utilizing different sections of the bristles (base, middle, and tip) based on their intended application. These products encompass shaving brushes, cosmetic brushes, brushes for coats and jackets, washing brushes, shoe brushes, carpet brushes, furniture dusting brushes, and combs for pet dogs.

Bristles sourced from local indigenous pigs demonstrate a considerably higher modulation index in comparison to those from exotic breeds such as Hampshire and Duroc. Nonetheless, the characteristics of bristles can differ based on the breed and the specific anatomical region from which they are harvested. Generally, bristles with diameters ranging from 210-320 μm contain 5-8 times more animal protein than their counterparts, like sheep wool, which has a diameter of merely 25-35 μm. The tensile strength of pig bristles is 4-5 times superior to that of human hair and other animal fibers, with local indigenous pigs yielding long bristles (5-7 inches) that possess a higher density of 150-188 bristles per square centimeter.

In comparison to synthetic options available in the market, brushes made from indigenous pig bristles are more durable, stable, and natural, resulting in a higher price point than other brushes found in Western nations. Furthermore, they provide enhanced flexibility, hardness, and effectiveness in removing dirt and dust, even from deep or zigzag corners. At present, while China and European countries manufacture various hair combs and cosmetic products from pig bristles, China remains the leading global producer of pig brushes.

Animal bristles, recognized for their distinctive properties and adaptability, have been utilized across multiple industries for centuries. Their inherent characteristics, combined with both traditional and contemporary processing methods, augment their value for a diverse range of applications. Grasping the complexities involved in their collection, processing, and application enables us to thoroughly recognize their importance in both common objects and specialized instruments.

References

Allden W.G. (2001). Feed intake, diet composition and wool growth. University of New England Publishing Unit, Armidale. 61-78.

Aten A., Innis R.F. and Knew E. (1955). Flaying and curing of hides and skins as a rural industry, FAO, Rome.

Biswas S. (2002). Monograph on "Meat Hygiene. WBUAFS, University Publication.

Black J.L. and Reis P.J. (2001). Speculation on the control of nutrient partition between wool growth and other body functions. Univ. New England Publishing Unit, Armidale. 269-294.

Booth J. E. (1996). Principles of Textile Testing. 3rd edn. New Delhi: CBS Publishers & Distributors

Dhantyagi S. (2012). Fundamentals of Textiles and their Care. 5th Ed. Hyderabad: Orient Longman Publisher.

Gracey I.F. and Collins, D.S. (1992). Meat Hygiene, 9th Edn.

Gracey J.F, Collin, D.S. and Huey, R.J. (2000). Meat Hygiene, 10th Edn, W.B. Saunders Company Ltd., London.

Handbook of Animal Husbandry. (2002). Indian Council of Agricultural Research Publication, New Delhi.

Hui Y.H., Nip Wai-Kit, Rogers R. W. and Young O.A. (Editors). (2001). Meat Science and Applications. Marcel Dekker, INC., New York.

Joseph M. L. (1987). Essentials of Textiles. USA: Holt Rinehart & Winston.

Kadolph S.J. (1998). Textiles. London: Prentice Hall.

Mann I. Processing and Utilization of Animal By-products. FAO agricultural Development Paper no. 75. FAO, UN.

Norman G. Marriot (1985). Principles of Food Sanitation. Van Nostrand Reinhold Company, New York.

Ockerman H.W. and Hansen C.L. (2000). Animal By-products Processing and Utilization. Lancaster, Publishing Company, USA.

Oddy V.H. and Annison E.F. (2000). Possible mechanisms by which physiological state influences the rate of wool growth. University of New England Publishing Unit, Armidale. 295-309.

Pizzuto J.J. (2011). Fabric Science. New York: Fairchild Publications,

Purser D.B. (2000). Effects of minerals upon wool growth University of New England Publishing Unit, Armidale. 243-255.

Sekhri S. (2011). Text Book of Fabric Science. New Delhi: PHI Learning Private Limited.

Sharma B.D. (2003). Modern Abattoir Practices and Animal Byproducts Technology, 1st Edn. Jaypee Brothers Medical Publishers (P) Ltd., New Delhi.

Tortora and Collier (1997). Understanding Textiles. London: Prentice- Hall,

Wynne A. (1997). Textiles, UK: Macmillan Education Ltd.

Index